QUANTUM LEAPS

How Quantum Mechanics Took Over Science

JEREMY BERNSTEIN

NEW JERSEY • LONDON • SINGAPORE • BEIJING • SHANGHAI • HONG KONG • TAIPEI • CHENNAI • TOKYO

Published by

World Scientific Publishing Co. Pte. Ltd.
5 Toh Tuck Link, Singapore 596224
USA office: 27 Warren Street, Suite 401-402, Hackensack, NJ 07601
UK office: 57 Shelton Street, Covent Garden, London WC2H 9HE

Library of Congress Cataloging-in-Publication Data
Names: Bernstein, Jeremy, 1929– author.
Title: Quantum leaps / Jeremy Bernstein (Stevens Institute of Technology, USA).
Description: Second edition. | Singapore ; Hackensack, NJ : World Scientific
Publishing Co. Pte. Ltd., [2018] | Includes bibliographical references.
Identifiers: LCCN 2018020505| ISBN 9789813272705 (hardcover ; alk. paper) |
ISBN 9813272708 (hardcover ; alk. paper)
Subjects: LCSH: Quantum theory--Miscellanea.
Classification: LCC QC174.13 .B47 2018 | DDC 530.12--dc23
LC record available at https://lccn.loc.gov/2018020505

British Library Cataloguing-in-Publication Data
A catalogue record for this book is available from the British Library.

For any available supplementary material, please visit
https://www.worldscientific.com/worldscibooks/10.1142/11056#t=suppl

Typeset by Diacritech Technologies Pvt. Ltd.
Chennai - 600106, India

Printed in Singapore

QUANTUM LEAPS

How Quantum Mechanics Took Over Science

Table of Contents

A Frontispiece

Whatever properties may ultimately be assigned to the atom, there is one which cannot be omitted—its power to seize and captivate the human mind. In fact, if we judged by the output of the printing press in the last few years, we might not unfairly assume that no sooner does anyone fall within the sphere of influence of this radiating personality, then he is seized with an irresistible determination to go home and write a book about it. Nor is the proselytising zeal confined to the pure physicist, whose protégé the atom may be presumed to be. We have books on the atom, some of them well done by chemists, by mathematicians, by technicians, and by journalists, and addressed to all sorts and conditions of readers. Thus we have "Atoms for Amateurs", "Atoms for Adepts", "Atoms for Adolescents", "Atoms for Archdeacons", "All about Atoms for Anybody"—these are not exact titles, but they indicate the scope of the volumes well enough—in fact, there seems to be a determination that no class of reader shall be left without an exposition of the subject suited to his condition and attainments. As these volumes continue to pour forth—we must assume that there are purchasers and readers. If we add to these the enormous output of serious scientific contributions from the many laboratories engaged in investigating the structures and properties of the atom, it is clear that this infinitesimal particle exerts an attraction unique in the history of science over the minds and imaginations of many types of men [and women]." *The Atom Again*, Nature, 1926, 118:365.

Acknowledgements

This is such an eclectic book and so many people have been kind enough to make suggestions and criticisms that I would like to distribute these acknowledgments by topics. On Auden I would like to thank Freeman Dyson, Nicholas Jenkins, Arthur Kirsch, Edward Mendelson, Oliver Sacks and Elizabeth Sifton. On the Dalai Lama I am grateful to Walter Isaacson and Abner Shimony. On all things Bohmian I am grateful to Ken Ford, Murray Gell-Mann, Basil Hiley and David Pines. On Léon Rosenfeld I am grateful to Loren Graham and Anja Skaar Jacobsen. On Newtonian matters I am grateful to Alan Shapiro and on Philipp Frank to Gerald Holton. On the vexed subject of the quantum theory of measurements I have had lively interactions with Elihu Abrahams, Steve Adler, Andy Cohen, Steve Gasiorowicz, Jim Hartle, Eugen Merzbacher and Bill Unruh. I have also had helpful comments from Peter Kaus, Arthur Miller, Bruce Rosenblum and Oliver Sacks. I would also like to acknowledge Michael Fisher of the Harvard University Press who kept the faith. I wish that I could acknowledge my debt to the late John Bell in person. His spirit hovers over this book.

An Introduction

> "For those who participated, it was a time of creation; there was terror as well as exaltation in their new insight. It will probably not be recorded very completely in history. As history, its re-creation would call for an art as high as the story of Oedipus of the story of Cromwell, yet in a realm of action so remote from our common experience that it is unlikely to be known to any poet or any historian."
>
> J.R.Oppenheimer, 1953 BBC Reith Lecture

When Oppenheimer made these Delphic comments in 1953, I was just beginning a serious study of the quantum theory to which he refers. Not that I had any broad knowledge, but I would have agreed with him that at that time quantum theory and its history were "unlikely to be known to any poet or historian". Things have changed. The last time I looked there were nine million three hundred and thirty thousand entries on Google under the rubric "quantum theory" and these certainly included both poets and historians as well as film critics and Buddhist monks. It would take a serious cultural historian, which I am not, to trace what happened in those fifty odd years. Having lived through it, while I have some theories, I am not sure that I can pinpoint exactly when this transformation occurred. I am however, quite sure that it was not that more people were learning quantum theory, but that more people were learning *about* quantum theory and especially about its interpretation and foundations. I am also quite sure that at the time Oppenheimer gave his lecture only a tiny number of physicists had any interest in the foundations. I will give an illustrative anecdote.

I spent two years at the Institute for Advanced Study in Princeton beginning in the fall of 1957. Oppenheimer was the director. We had a weekly seminar which he attended. By the spring most of the obvious speakers had spoken and it was getting more difficult to recruit new ones. A young and gifted colleague of mine was recruited on the basis of a preprint he had written on the quantum theory of measurement, to this day a highly controversial subject. Physicists disagree on how measurements, which are described by classical physics, fit into quantum theory. It was all but unheard of for a young physicist to work on something like this. It was not even considered physics. I admired his courage. I do not think that he asked to give a seminar. I think he was told to. Oppenheimer was notorious for cutting down seminar speakers whom he thought were wasting his time. At this seminar he outdid himself. The speaker had gotten out about five sentences when Oppenheimer said that Niels Bohr had answered all these questions in the 1930s and the speaker had nothing to add. It was the end of the seminar. I am glad to report that my colleague published his paper and that he went on to have a distinguished career.

I contrast this with another scene I witnessed in the 1980s—thirty odd years later. This took place at Columbia University. John Bell, whom I think is largely responsible for the renewed interest in this subject, had been invited to give a colloquium. This was meant to address an audience consisting of the entire physics department. It was held in the largest lecture room available to the department. It was packed-standing room only. People came from every university in New York city and its surroundings—even some from New Jersey. I can assure you that no one stopped Bell to tell him that Bohr had done it all.

The purpose of this book is to give an account of this cultural transformation. The subject matter is eclectic ranging from the Dalai Lama to W.H.Auden. I try to explain the relevant parts of the theory as I go along. There is essentially no mathematics. Words cannot really replace mathematics, but in a subject like this mathematics cannot really replace words. There is a good deal of autobiography here. I hope the reader will not find this intrusive but that is the kind of writer I am. I am sure that I have not covered the entire subject. There is too much for any one person to cover. But I hope that I have covered enough to make it clear that this subject is now known to both poets and historians as well as playwrights, novelists and film-makers, Buddhist monks and communist ideologues, as well as physicists.

Chapter 1

Bishops

"Rhyme-royal's difficult enough to play.
But if no classics as in Chaucer's day,
At least my modern pieces shall be cheery
Like English bishops on the Quantum Theory."

W.H.Auden, Letter to Lord Byron, 1936

"Ah, my Lord of Birmingham, come in, sit on the fire and anticipate the judgement of the Universal Church."[1]

Right Reverend Herbert Hensley

"I like the company of men of science: they are not excessively intellectual in their hours of leisure and they give good dinners."

Bishop of Birmingham E.W.Barnes[2]

In the fall of 1957, I began what turned out to be a two-year membership at the Institute for Advanced Study in Princeton. This was a banner time for physics. In 1956, the Chinese-born American physicists T.D.Lee and C.N.Yang noted in several papers that what is called "parity symmetry"—the symmetry between left and right handed descriptions—a cornerstone of physics up to that point—had never been tested in the sort of weak interactions that are responsible for the instability of many elementary particles and atomic nuclei. They proposed several experiments, and when these were performed they demonstrated that, in these interactions, the symmetry was indeed violated. This was a sensational discovery and all of

[1]**This quotation is taken from John Barnes, *Ahead of His Age; Bishop Barnes of Birmingham*, Collins, London, 1979. p.190. I shall have occasion to make several references to this book which I shall call "AA".**

[2]**AA, 344.**

us were gripped by its implications. In the fall of 1957, Lee and Yang, who were both at the Institute, were awarded the Nobel Prize for Physics. Soon after my arrival at the Institute I studied the membership roster. There were several physicists whose work I had studied and admired, but I came across a name that really surprised me—Reinhold Niebuhr.

Niebuhr, whom I had never met, was a hero of mine. During the ten years I was at Harvard (1947–1957), I had heard him preach and debate several times. He was a great orator. His language was simple but his ideas were profound and complex. He was also an imposing figure and gave the impression that you were directly in his gaze. When he spoke of the dangers of neurotic preoccupation with self, I was sure that he was talking specifically to me. As it happened, my father, who was a rabbi, had been on several liberal commissions with Niebuhr, so I thought that if the occasion presented itself I would introduce myself to him. In the event, the occasion presented itself quite quickly. We all took our lunches in the Institute cafeteria and I noticed Niebuhr and his wife eating alone. I passed by their table and said hello. They invited me to sit down. When Mrs. Niebuhr—Ursula—and I were alone for a minute she told me that her husband had suffered a series of strokes, which was one of the reasons he had sought the relative tranquility of the Institute. She said that he was feeling a little depressed and that if I would drop by their apartment from time to time it might help to cheer him up. I took advantage of this invitation. Niebuhr had sensed the excitement of the physicists and wanted me to explain what they were so excited about, which I did as best I could.

On one visit I noticed a book by W.H.Auden. I now forget which. Auden was another of my heroes and, as it happened, he was giving the Christian Gauss Seminars on literary criticism at the Princeton University, and I had attended some of his lectures. I mentioned this to the Niebuhrs and they were both surprised and delighted. I will never forget how Ursula Niebuhr pronounced Auden's first name—Wystan. It sounded like "whistle"—with a hiss on the 's.' I did not know at the time that the Niebuhrs and Auden were close friends. Although she and Auden had been at Oxford at the same time—the late 1920s—they had only met in 1940. Niebuhr had become a sort of spiritual advisor to Auden who indeed dedicated his collection, "Nones", to them. I felt that I had conveyed a useful piece of information to the Neibuhrs and forgot about it.

However, not long afterwards I was taking the "dinky"—the small train that connects Princeton to Princeton Junction—when on came Auden. You couldn't miss him. He had, at the time a face, which he said looked like a wedding cake that had been left out in the rain. The train was pretty full but there was an empty seat next to me and Auden took it. The ride was short but I told him that the Niebuhrs were in Princeton, which he had not known. He seemed pleased to learn this and again I more or less forgot about it. But some days later in the morning I was called to the hall telephone in our building. None of us had private phones in our offices since Robert Oppenheimer, our director, had decided that these might distract us from our work. The hall phone distracted everyone from their work. It was a call from Oppenheimer's secretary. She said that I was expected for lunch and should come at lunch time to Oppenheimer's office. I replied that there was surely some mistake since there was no earthly reason why Oppenheimer would want to have lunch with me. She said there was no mistake and I was expected. Even as I write this over a half century later I can still see the scene in Oppenheimer's office. There was Oppenheimer, impeccably dressed as usual, and his wife "Kitty." There was Sir Llewellyn Woodward, a notable British historian who had come from Oxford to the Institute, and his wife. There were the Niebuhrs, and Auden, and myself, a very minor post-doctoral. Whatever this was about, I am sure it reflected the fine "Italian hand" of Ursula Neibuhr, who wanted Oppenheimer and Auden to meet. After a few introductions we all marched into the cafeteria.

A special table had been prepared for us in the center of the dining area. I am sure that we were served and did not stand in the cafeteria line. There was probably also wine. I remember the very fishy looks that my colleagues who were at our usual tables gave me. Freeman Dyson, who was then a professor at the Institute and a friend of mine, looked particularly amused. I had no opportunity to explain. I wish I could tell you that the conversation was memorable. Oppenheimer, who was sitting across from Auden, seemed rather ill at ease. At one point he told Auden that he had studied Sanskrit in Berkeley in the 1930s. This did not make any impression and Auden and Ursula Niebuhr then engaged in a lively conversation which pretty much ignored the rest of us. The Woodwards said nothing. Niebuhr caught my eye, and from the look on his face I would guess he was thinking that this too would pass. After lunch, I took Auden to meet Dyson. They played some word games on Dyson's blackboard, and then Auden left and I never saw him again.

I have thought about this extraordinary occasion many times over the years, usually with a sense of regret. If only I had it to do again, I would have asked Auden which "English bishops on the quantum theory" he was referring to in his Byron poem. What a conversation that might have turned into. But, myself excepted, everyone who was at that table is now gone. "English bishops on the quantum theory," how many can there have been? Finally, I decided to try to find out. The first person I asked was Dyson. His answer was immediate, the Bishop of Birmingham, Ernest William Barnes. This suggestion was reinforced by the Auden scholars Edward Mendelson and Arthur Kirsch. But I have to confess that after Dyson said his name I did not have the foggiest idea who Bishop Barnes was and why he would have had anything to say about the quantum theory. I am now going to tell you what I have learned.

Barnes was born on the 1st of April, 1874—All Fools' Day—in the town of Altrincham in Cheshire. His father Starkie (John) Barnes was a schoolmaster and his mother Jane née Kerry was the daughter of the village shoemaker at Charlbury in Oxfordshire. But in 1876 the family moved to Birmingham. I would be fascinated to know at what age and in what way Barnes began to show special mathematical ability. Mathematicians begin very young. To take an example, Dyson once told me that when he was still young enough to be "put down for naps" he invented for himself the notion of the convergent infinite series. He noticed that if you added $1+1/2+1/4+1/8 \ldots$, the sum approached two. Where in this spectrum was Barnes? I do not know. In 1886, Barnes entered the King Edward VI Grammar school on a scholarship. Secondary education then was neither free nor compulsory. All of Barnes' higher education was done on scholarships. The family could not have afforded the tuition. At King Edward's, Barnes had the good fortune to come upon a truly great mathematics teacher, Rawdon Levett, who recognized Barnes' ability and was able to guide him into learning things like non-Euclidean geometry. Barnes was grateful to Levett for the rest of his life. It was Levett who encouraged Barnes to apply for a fellowship to Trinity College in Cambridge, which he won.

To graduate with honors from Cambridge, the student had to take the so-called Tripos examination. The mathematics Tripos consisted of hard problems that the student had to complete in a certain time limit. The student who did best was labeled the Senior Wrangler. Where the term "wrangler" came from in this context, no one seems to know. The student who did worst was known as the "Wooden Spoon" and was indeed given a spoon along with third class honors. As one might imagine, the list of Senior Wranglers from that period is quite impressive. It includes such people as the physicist Lord Rayleigh and the astronomer Arthur Eddington. But what one might not expect is that the list of Second Wranglers is even more impressive. It includes such people as James Clerk Maxwell, the greatest physicist of the 19th century, and William Thomson—Lord Kelvin—who was not far behind. Thomson's case is instructive. In one of his courses he had come in with a theorem and its proof. The theorem was on the Tripos and Kelvin had forgotten the proof. He spent a good deal of his time reconstructing it. The man who became the Senior Wrangler had it memorized. One wonders how Einstein would have done on the Tripos.

In 1896, Barnes was Second Wrangler. The Senior Wrangler was a New Zealand-born physicist named Richard C. Maclaurin, hardly a household name although he did become the president of MIT. But two years later Barnes won the First Smith's Prize for his Ph.D. thesis, an even greater honor which led to a fellowship at Trinity. For something like a decade, Barnes did very significant work in pure mathematics. His work in the theory of functions is still being adumbrated. When I looked it up on the web, I was interested to note that one of the papers cited was written by Dyson. In 1909, at the age of thirty five, Barnes was elected to a Fellowship in the Royal Society. Soon after, he stopped doing creative mathematics. Dyson told me that G.H.Hardy, one of the most distinguished of the 20th century mathematicians, told him that once Barnes had been elected to the Royal Society he dropped mathematics like a "hot brick." It would certainly seem that way. Perhaps Barnes had realized that he would never be a mathematician in the class of people like Hardy and above all Ramanujan. Nobody was in the class of Ramanujan. He had been discovered by Hardy in 1913, who had invited him to Cambridge. He needed a formal tutor, and the more senior Barnes took on the responsibility. Barnes was remembered as a very good lecturer with a deep interest in mathematical education. Maybe it was not enough.

Barnes's family's religious tradition was Baptist and Wesleyan, but the King Edward School was strictly Church of England. Barnes had no difficulty adapting and his parents did not object. In 1902 Barnes became ordained as a deacon in the church, and from that time on preached sermons in various locales around Cambridge and elsewhere. From the beginning, his sermons had scientific content. Confronted with the facts of evolution, for example, he could not take the bible as a literal cosmogony. These sermons were referred to by others as Barnes's "gorilla" sermons. Another consistent theme of Barnes's sermons was pacifism. He was a confirmed pacifist from the time of his youth, and he never changed through two World Wars. At the time of the First World War this got him in trouble with his more conservative colleagues, though nothing like the trouble Bertrand Russell got into. Russell was then at Cambridge and his vivid denunciations of the War landed him in jail. Barnes was prepared to give Russell a character reference even though they were not close friends and Barnes did not much approve of Russell's lifestyle. In 1915, Barnes received an offer he could not refuse. The Temple Church in London was looking for a new "Master." The church, which was constructed by the Knights Templar in the 12th century to serve their spiritual needs, had in the 20th century a very distinguished congregation. Barnes was only forty three—young to be a Master of the Temple. It meant leaving Cambridge and moving to London. There was also a personal matter. Barnes had fallen in love with Adelaide Ward, the only daughter of a distinguished family. She had great hesitations about marriage but relented when Barnes told her that he would refuse the job unless she came to London with him. It was a very happy marriage.

Barnes remained at the Temple until 1924. Then he received a letter from Prime Minister Ramsay MacDonald informing him that his name had been submitted to the King for an appointment as the Lord Bishop of Birmingham. The Church of England was the state religion and such appointments required royal approval, which Barnes got. It automatically carried a membership in the House of Lords. This enabled Barnes to participate in the debate in Parliament in 1927–28 about revising the Book of Common Prayer, which could only be done with the consent of Parliament. Barnes opposed the revisions and they were defeated by a narrow margin. When it came to church practice, Barnes was known to be very strict about not allowing any trace of Catholic ritual to enter into Anglican practice. And on the matter of transubstantiation he was dogmatic. In a sermon he said, " I am quite prepared to believe in transubstantiation when I can find a person who will come to the chapel of my house and tell me correctly whether a piece of bread which I present to him has undergone the change for which believers in transubstatiation contend ..."[3] As one might imagine, this sort of thing rendered Anglo-Catholics, of which Auden's mother was one, livid. That Auden would make a reference to Barnes—however concealed—in his poem must tell us something about his relationship to his mother. In fact this was not Auden's first mention of Barnes in his poetry. The earlier one is much more explicit and curious.

In September of 1931, the British Association for the Advancement of Science had its centenary meeting. It was a stellar gathering featuring such things as a lecture on "holism", his term, by General Jan Christaan Smuts. There were the public scientists of the time, such as James Jeans. Jeans

[3]**AA, 194.**

had been a first rate physicist, but by the 1930s he was writing popular books with a Christian mystic bent. Barnes, who was also there, must have found this appealing. This meeting must have caught Auden's attention. There seem to be echoes in his poem, The Orators, which was written soon after the meeting.[4] It is a remarkable and strange poem—partly in prose and partly in rhyming verse. Towards the end come the lines,

> Their day is over, they shall decorate the Zoo,
> With Professor Jeans and Bishop Barnes at 2d a view,
> Or be ducked in a gletcher, as they ought to be,
> With the Simonites, the Moselyites and the I.L.P.

"Their" refers to various people that Auden knew. But what an odd association, to include Jeans and Barnes and the Simonites, Mosleyites and the Independent Labour Party.

The Gifford lectures, well-known among theologians and delivered every year at one or another Scottish university thanks to a bequest of Adam Lord Gifford who died in 1887. were intended to "promote and diffuse the Study of Natural Theology in the widest sense of the term ... in other words, the knowledge of God." "Natural Theology" meant theology consistent with the established principle of science. Over the years the Gifford lecturers have been an extraordinary group. The scientists included people like Niels Bohr and Werner Heisenberg. There were philosophers such as Henri Bergson and Albert Schweitzer. Niebuhr gave the 1938–1940 series. Dyson told me that when he gave the 1984–1985 series in Aberdeen no one was allowed to ask questions. Apparently you were allowed to laugh. In his biography of his father, John Barnes notes that when his father gave the 1927–1929 series he provoked laughter by filling the blackboard with equations from Einstein's theory of general relativity. I would imagine that it was nervous laughter. The lecturers were expected to produce short books that contained their lectures which were published not too long after them. Barnes' book, *Scientific Theory and Religion,*[5] his six-hundred-and-eighty-five-page *magnum opus*, was not published until 1934. He spent every moment of his spare time in the intervening years preparing it.

---

[4]**I am grateful to Edward Mendelson for this information and for other insights into the poem.**

[5]**John Barnes, *Scientific Theory and Religion,* Cambridge University Press, Cambridge, 1934. I will refer to this as ST.**

The book covers all the natural sciences. There is rather little about mathematics. Perhaps Barnes did not think of it as a natural science. From time to time he can't resist a dig. For example he writes, "I have, personally, little doubt that biological research will, in due course, prove a human virgin birth is possible."[6] He adds,

> "There are some who may shrink from any such investigation of the Virgin Birth as I have indicated. For my own part I am convinced that we must abandon the practice of arguing in vacuo, as it were, for such a mode of argument was the typical advice of medieval scholasticism. On the basis of observed facts of Nature, and by arguments drawn from analogies as come within the range of our observations, we must approach all problems which present themselves. I would add that reverence and truth can always be combined, unless the object of our reverence happens to be untrue." *Pace* Auden's mother.

Barnes' chapters on the theory of relativity are really excellent. I doubt that many professional physicists could have written better ones. I was particularly interested in what he had to say about cosmology. The modern era in cosmology began when Einstein published a paper on the subject in 1917. This was a year after he had published his masterpiece on general relativity and gravitation. At the time the "cosmos" consisted only of our own galaxy, and Einstein's concern was that his theory predicted that the gravitational attractions would make the whole thing collapse. So in his 1917 paper he added a kind of anti-gravitational term to his original equations to keep the universe stationary. However, a few years later a Russian, Alexander Alexandrovich Friedman, completely unknown to Einstein, sent him a copy of a paper he was going to publish which showed that, depending on how the matter was distributed in the cosmos, there were solutions of Einstein's original equations in which the universe expanded or contracted. Einstein's first reaction was to claim that Friedman had made a mistake. But it was Einstein who had made a mistake. Then he claimed that Friedman's solutions were correct but irrelevant since the universe was stationary. Here things stood until 1927.

---

[6]**ST, 458.**

At this time, a Belgian Catholic priest and astronomer named Georges Henri Joseph Édward Lemaître published a paper in which he had rediscovered Friedman's solutions. But he made a crucial next step. If the universe is expanding according to these equations, then every galaxy is receding from every other galaxy at a speed that increases to first approximation as the distance of separation increases. We can think of the galaxies as being on the surface of a balloon that is being blown up. All the dots on the surface are receding from all the other dots. But Lemaître argued that this had an observable consequence. The spectral colors of the light from these galaxies should be shifted to the red—the Doppler shift—as the distance, and hence the speed, of recession increases. In 1929, the American astronomer Edwin Hubble published his observational results that showed that this indeed is what is happening. In his paper, he says that he is confirming the prediction of Lemaître. Einstein gave up his objections. But ironically, it now looks as if the universe is expanding too rapidly to be accounted for by the Friedman equations and some "anti-gravity" is needed.

Barnes discusses the Lemaître model in detail. He concludes that there is something wrong. The time scales don't work out. If the universe is some millions of years old the observed distances would have to increase too rapidly. But, in fact, cosmologists now think that the age of the universe is about 13.7 billion years so that the time scales do work out. Lemaître discusses where it all came from. He envisions a "cosmic egg" that exploded and voilà tout—the first Big Bang model. But where in this is God? Here is what Barnes writes,

> "Must we then postulate Divine intervention? Are we to bring in God to create the first current of Laplace's nebula or to let off the cosmic firework of Lemaître's imagination? I confess an unwillingness to bring God in this way upon the scene. The circumstances which thus seem to demand His presence are too remote and too obscure to afford me any true satisfaction. Men have thought to find God at the special creation of their own species, or active when mind or life first appeared on earth."
>
> "They have made him God of the gaps in human knowledge. To me the God of the trigger is as little satisfying as the God of the gaps. It is because throughout the physical

Universe I find thought and plan and power that behind it I see God as the creator."[7]

This does not differ a great deal from Einstein's "old one"—his God.

Barnes' chapter on quantum theory, unlike the ones on relativity, seems a bit antiquated to me. To give an example, after his discovery of the atomic nucleus with his young collaborators, Hans Geiger and Ernest Marsden, the New Zealand-born physicist Ernest Rutherford proposed a model of matter. The nucleus had a positive electric charge to balance the negative charges of the surrounding electrons to keep the atom electrically neutral. These positive charges were provided by the protons in the nucleus. But there had also had to be neutral objects of about the same mass as the protons to make the masses of the nuclei come out. Rutherford made the natural assumption that these neutral objects were electrons and protons bound together. Indeed, when one of Rutherford's students, James Chadwick, made a direct observation of these neutral objects in 1932, he announced that this is what he had discovered. Already there were naysayers such as Wolfgang Pauli who said that this must be wrong. But Chadwick dismissed them. He was wrong and, by 1934, when Barnes published his book, there was no doubt that these neutral objects—'neutrons' as they were named—were elementary particles in their own right. Barnes describes Chadwick's experiment, but he presents Rutherford's old model as if it were still valid. I find this very puzzling and perhaps reflective of the fact that Barnes was not much in contact with contemporary physicists. This manifests itself in the rest of the chapter.

As far as I could see there is nothing wrong in what he wrote. It is what he didn't write that struck me. He mentions, almost in passing, the work of Erwin Schrödinger, but then does not exhibit his equation, which is the heart of the matter. The mathematics is much simpler than that of general relativity, which he goes into in considerable detail. He also mentions, in passing, the work of Paul Dirac, but never mentions Dirac's 1928 discovery of the equation that bears his name, which brought relativity and the quantum theory together and which predicted the existence of anti-particles. The first of these, the positron, was discovered in 1932, two years before Barnes published his book. He has a clear, brief discussion of the Heisenberg uncertainty principle, which Heisenberg presented in 1927. But there is no hint of the debate between Einstein and Bohr on the meaning of the theory, which

---

[7]**ST, 409.**

had begun about the same time. It is a strange chapter. I kept wondering: Did Auden read it? Was this his "Bishop cheery" writing on the quantum theory? Perhaps Auden read one of the concluding paragraphs.

> "Old traditions embedded in Christian thought, describe a time when 'the earth was without form and void': and probably most Christian philosophers in the past believed in a process of creation by God, in time and 'out of nothing'. We seem, in the analysis of matter to which Einstein's general relativity leads, to see 'in the beginning' a process by which form or structure was given to the void of space-time. As the many complex forms which then arose assumed an ever greater complexity, the material world took shape. It is natural to ask whether, in such development, there was creative activity, the emergence of something new. I feel constrained to answer in the affirmative. Things were other and more diverse at the end than at the beginning. Even though the process of assuming greater complexity may be exhibited by a mechanical sequence, it may none the less conceal or embody genuine creative activity ... Did not such complexity involve in its making a series of acts rightly to be called creative, or at least a series of changes which led to the creation of something that did not previously exist?"[8]

As for Barnes, in 1947 he published a very controversial book, *The Rise of Christianity*, in which he expressed his rationalistic views fully. It created a furor with attempts, unsuccessful, to make him resign. He finally did in February of 1953. He died in November that same year.

In 1940, the chocolate magnate, Edward Cadbury, established a chair in theology at the University of Birmingham.

Niebuhr was in the process of giving his Gifford lectures and Barnes proposed him as the first occupant of the Chair. The suggestion was turned down. But let us suppose it had not been. Then the Niebuhrs would not have met Auden and I would not have met any of them. The encounter that inspired me to make this inquiry in the first place would never have happened.

[8] **ST, 191–2.**

Chapter 2

Quantum Buddhists

"On the Quantum Cruise you will directly experience your connection with the Unified Field through music, movement, art, sound, drumming, Quantum Imagery, and other dynamic, fun and transformative processes. Integrate your intellectual understandings into the very molecules of your physical body."

The Quantum Cruise Website

"My confidence in venturing into science lies in my basic belief that as in science so in Buddhism, understanding the nature of reality is pursued by means of critical investigation: if scientific analysis were conclusively to demonstrate certain claims in Buddhism to be false, then we must accept the findings of science and abandon those claims."

The Dalai Lama[1]

"When one listens to descriptions of subatomic particles such as quarks and leptons in modern physics, it is evident that the early Buddhist atomic theories and their conception of the smallest indivisible particles of matter are at best crude models. However, the basic thrust of the Buddhist theorists, that even the smallest constituents of matter must be understood as composites, appears to have been on the right track."

The Dalai Lama[2]

On August 30, 1983, the Dalai Lama and a delegation of Tibetan monks visited CERN. "CERN" is an acronym for the name 'Conseil Europé en

[1]**The Dalai Lama, *The Universe in a Single Atom*, Morgan Road Books, New York, 2005. 2 – 3. This statement by the Dalai Lama does not make it clear that these particles have to be created in what is called an "entangled" state. This is further explained in chapter 6. Dalai Lama, 64 – 6.**

[2]**Dalai Lama, 55 – 6. Dalai Lama, 64 – 6.**

pour la Recherche Nuclé aire' although in the official name, the word 'Organisation' has now replaced 'Conseil'. CERN, which is on the border between France and Switzerland just outside of Geneva, is the largest laboratory in the world devoted to the physics of elementary particles. It is an international organization in every sense, although I am not aware of any Tibetans who have ever worked there, at least as scientists. Why the Dalai Lama chose to make this visit is somewhat mysterious to me, although accounts of his boyhood describe his interest in taking apart and putting together machines to see how they worked. In any event, on this occasion there were something like thirteen Tibetan monks. CERN, which is used to receiving delegations, arranged a formal lunch for the Tibetans and a corresponding number of people from the laboratory. One of them was my friend, the late John Bell who died suddenly of a cerebral hemorrhage in the fall of 1990 at the age of sixty-two. But some months before Bell's death, I had had the chance to discuss the Dalai Lama's visit at some length with him.

The reason for Bell's having been part of the CERN delegation is quite clear. Bell had almost single-handedly renewed interest among physicists in the foundations of the quantum theory. Until Bell's work, a few decades before this visit, this subject was pretty much the province of elderly physicists such as Bohr and Einstein, with very few others paying much attention. The Einstein–Bohr debates did not seem to lead anywhere. Bell's work, which I will discuss more fully later, made the subject an experimental science. It received wide publicity and considerable misunderstanding ensued, which I will also discuss later. There was, and is, an attempt to link it with Eastern religions such as Buddhism, and this was possibly one of the motivations that led the Dalai Lama to want to visit CERN. In any event it was a "state visit" in which the Dalai Lama, who speaks English both fluently and charmingly, would only speak through an interpreter. Bell recalled some of the discussion.

At one point the Dalai Lama asked how the elementary particles could add up to make extended objects if they were point-like. "It did not seem to occur to him that there could be spaces between the points—very odd," Bell had remarked. Bell was interested in exploring with the Dalai Lama the notion of the Big Bang. He cautioned that the concept of the Big Bang was only a fad in modern science which could change. But could it be reconciled with the Buddhist canon, according to which the universe is constantly recycled? The Dalai Lama said that this was not part of the canons to which they were completely committed, but they would have to study the scriptures carefully. Then he added, "There is some room for maneuver." I was so struck by the "some room for maneuver" that I asked Bell whether this was his own turn of phrase or that of the Dalai Lama's. Bell said that the phrase "room for maneuver" is exactly what the interpreter said, and Bell found that wonderful. In fact, although Bell did not tell me this—he may have forgotten or never knew—this session was filmed and one can see both Bell asking this question and the Dalai Lama answering.[3]

Bell brought up the question which is often asked—why science did not take root in countries like China with a Buddhist tradition.[4] China had had a long history of inventions—many of them of great significance. But nowhere do you find a Galileo, a Newton, or a Darwin, to say nothing of an Einstein. There must certainly have been potential scientific genius, but there was no context. A tiny country like Denmark could produce a Bohr, because once such a person appeared, there was a tradition into which he could be fitted. Mathematics is a little different. A genius can almost make his own context. Ramanujan, for example, had some mathematical education, but when he went to Cambridge to work with G.H. Hardy, he did not know what a proof was. He simply produced fantastic results out of his imagination, that he knew were true. Bell noted, "Buddhism is an inward religion concerned with individual salvation. The study of how stones accelerate must have seemed unimportant. The Tibetans were very hazy about modern physics and I had the impression that they did not consider these analogies significant. Physics is ephemeral and rapidly changing and applies to only a limited set of phenomena. It would be absurd to anchor your religious beliefs to theories of physics. I did not have the impression that they were looking around and grasping for straws. They seemed to be quite comfortable in their own traditions."

I took advantage of my interviews with Bell to ask him about another encounter he had had with Eastern religious mysticism. This one took place in 1979 at a conference with the Maharishi Mahesh Yogi, who died in February of 2008. As it happens, the Maharishi had been a physics major at the University of Allahabad in the North Indian state of Uttar Pradesh. In 1939, after his graduation, he became a disciple of Swami Brahmananda Saraswatī who gave him the name Bal Brahmachayra Mahesh. Not being a Brahmin, the Maharishi could not succeed his master after the latter's

---

[3]**I am grateful to Luis Alvarez Gaumé and Pierre Jarron of CERN for providing me with a CD of this film.**

[4]**In the film Bell actually reads a quote from Joseph Needham to this effect.**

death in 1953. So he struck out on his own, teaching Transcendental Meditation to various disciples including, most notably, the Beatles. In the early 1970s he founded the eponymous Maharishi European Research University in what had been an elegant resort hotel near Weggis, above Lake Lucerne in Switzerland. The hotel had a large circular dome which bore inscriptions such as "Center of World Government." There was a Minister of Health and Immortality. He appointed as Chancellor a physicist named Larry Domash.

A brief Google search turns up a 1973 paper entitled "Interactions among Multiple Lines in the 8446-A Atomic-Oxygen Laser" published in *the Physical Review*—our refereed trade journal—co-authored by Domash. The byline indicates that, when the paper was written, he was employed by the NASA Electronics Research Center, in Cambridge, Massachusetts. But there is a footnote that reads "Present address, Maharishi International University, Rishikesh, India." It was surely Domash who selected the list of invitees for the 1979 conference. In addition to Bell, there were other well-known quantum theorists. "The Maharishi," Bell recalled, "was a much more regal figure than the Dalai Lama. He sat on a sort of throne dressed in white robes. At his feet were acolytes—young women—also dressed in white robes. In the middle of this gathering were the invitees. It was a very uncomfortable situation for a scientist. I made a speech that was skeptical in character. Domash was trying to see some analogy between the state which people reach in meditation and the ground state of a superconductor." A "superconductor" is a quantum mechanical state that some materials attain when they are cooled to very low temperatures. As the name implies, electricity is conducted with no resistance in this state. I have read claims by Domash that neurons are superconductors. Bell thought that all of this was nonsense and said so.

"They asked the Maharishi for his views. I was shocked to learn that the Maharishi claimed he could make rain. First you see a blue sky and a little cloud and then you relax. The cloud grows and there is rain. Unfortunately this was not something you could order, especially with a skeptical audience. At that time he had a class whom he was trying to teach to levitate above the ground. He was interested in all the queer aspects of quantum mechanics which he thought had some connection to Eastern mysticism. I did not have a chance to have a conversation with him. For me he was just a figure on the throne making pronouncements. But I liked the Maharishi setup. The meals were good. It was vegetarian." Bell had been a vegetarian since the age of sixteen.

While Bell had a good deal of skepticism about the Maharishi, his feelings for the Dalai Lama were quite different. Knowing John Bell as I did, I am sure that his description of his meeting with the Dalai Lama is an accurate representation of what occurred. In fact, it is born out by the film. The pity is that Bell and the rest of the CERN people did not know what the Dalai Lama's scientific background even was in 1983. If they had, or if Bell had had a chance to talk to the Dalai Lama privately and informally, I think the results would have been very different and the impressions created would have been very different. As it was, Bell gave a short, but brilliant, introduction to the quantum theory. There was an attempt at dialogue which foundered because of all the people present. In 2005, the Dalai Lama published a small book entitled *The Universe in a Single Atom: The Convergence of Science and Spirituality.*[5] In the book, he makes it clear that his interest in science goes back to his early boyhood. The 13th Dalai Lama—the "previous body"—died in 1933. The search for his reincarnate did not begin for four years. In 1937, a search party from Lhassa found a two-year-old boy in Amdo—a province in the North—whom they recognized as the incarnate—the "present body", as the Dalai Lama refers to himself. They had to pay a large bribe to the Chinese, who claimed sovereignty, to allow the child to leave. Once in Lhassa, he was installed in the Potala Palace and his education in Buddhism began. Science played no part in it and I am not sure he was taught even geometry.

The Potala is a giant warren which has, it is said, a thousand rooms. It is now, as I can testify having first visited it in 1987, a museum in which the Chinese extract as much money as they can from tourists. Tibetans can't afford to go there. Among the exhibits are the stupas, the gold encrusted tombs in which all the Dalai Lamas from the Fifth to the Thirteenth lie in state. The Thirteenth was an interesting man. He tried without success to modernize Tibet. During the early years of his regime there was a power struggle between the Chinese and the British as to which country had domain over Tibet. The Chinese invaded in 1910, and the Dalai Lama fled to Sikkim, a Himalayan British protectorate. He used the same route that his successor would use in 1959. The political officer in charge of Sikkim at the time was a man named Charles Bell—no relation to John.

Charles Bell and the 13th Dalai Lama had an amiable relationship and he was given several "souvenirs" of his visit. There was a telescope and a hand wound pocket watch which gave the time in various time zones.

[5] **Op cit.**

There were also three automobiles acquired later—two 1927 Baby Austins and a 1931 American Dodge. There was—and still is—no motorable road from Sikkim to Lhassa, so the automobiles had to be dismantled and the parts carried over the Himalayas by porters, human and animal. They were re-assembled and maintained, but how much they were driven around the city—the only place in the country that had anything like roads—is unclear. The young fourteenth Dalai Lama, while exploring the Potala, discovered these treasures. He managed to take a short drive in one of the cars, and had a small accident in which a headlight was damaged. The watch he learned to take apart and put together. No one explained to him the use of the telescope, but he figured that out too. He began observing the Moon and noticed that there were shadows being cast by various protuberances. He called in his tutors who did not understand what they were seeing. The Dalai Lama explained it must mean that the Moon was being illuminated by the Sun—a scientific discovery. Using his telescope he made another discovery—there was a European in Lhassa. This turned out to be the Austrian mountain climber Heinrich Harrer.

In 1939, Harrar was part of a German expedition to climb Nanga Parbat. This is the ninth highest mountain in the world and one of the most difficult to climb. It is located in Pakistan which was then part of British India. The war broke out and the climbers were interned as enemy aliens by the British. It is very unlikely that the British knew that Harrer was in the SS, and had been photographed with Hitler. He was a German hero for having climbed the north face of the Eiger in Switzerland. The Dalai Lama certainly did not know that Harrer had been in the SS. It was only revealed publicly in 1997. Harrer's attitude was that bygones should be bygones. He died in 2006 at the age of 93. Be all that as it may, Harrer, who had escaped from the British internment camp in 1944, and arrived in Lhassa after a harrowing trek across Tibet in 1946, was the Dalai Lama's first contact with the outside world. He repaired the movie projector and the two of them watched together such films as Lawrence Olivier's *Henry V*, which had been imported from India. Harrer also taught the Dalai Lama world geography, something that was new to him. Harrer had been accompanied on his escape by another Austrian mountaineer named Peter Aufschnaiter, who was actually the leader of the Nanga Parbat expedition. While in Tibet, he mapped very remote areas of the country for the government and helped design the hydroelectric power plant in Lhassa. After the Chinese invasion in 1950, both men left the country, Harrer for Austria and Aufschnaiter for Nepal. The latter had married a Tibetan woman. I met Aufschnaiter and his wife when I was in Nepal in 1967. He was doing agricultural engineering for various international organizations, He died in 1973. In 1953, Harrar published his much admired *Seven Years in Tibet* and this was followed by Aufschnaiter's *Eight Years in Tibet.* In October of 1950, the Chinese army invaded Tibet. Not long afterwards the Dalai Lama, at age sixteen, was invested in the full powers of his office, something which would not have happened ordinarily until he was eighteen. During this period he visited China and met Mao. In his book, the Dalai Lama says that this was the first time he had ever seen a real city with all its technological variety. He also visited India where he was to flee in 1959. It was in 1973, when he made his first visit to the West. In England, he does not tell us how, he met the Austrian-born philosopher of science, Karl Popper. He and Popper shared the experience of having been exiled forcibly from their native countries. The Dalai Lama says that his English was not yet good enough to get the full benefit of his conversations with Popper. Popper's main theme in the philosophy of science was the notion of "falsifiability." No number of confirmations suffices to prove a theory, but one example of a prediction that fails, suffices to disprove it. The Dalai Lama writes,

> "Another of the differences between science and Buddhism as I see them lies in what constitutes a valid hypothesis. Here too Popper's delineation of the scope of a strictly scientific question represents a great insight. This is the Popperian falsifiability thesis, which states that any scientific theory must contain within it conditions under which it can be shown to be false. For example the theory that God created the world can never be a scientific one because it cannot contain an explanation of the conditions under which the theory could be proven false. If we take this criterion seriously, then many questions that pertain to our human existence, such as ethics, aesthetics, and spirituality, remain outside the domain of science. By contrast, the domain of inquiry in Buddhism is not limited to the objective. It also encompasses the subjective world of experience, as well as the question of values. In other words, science deals with empirical facts but not with metaphysics and ethics, whereas for Buddhism, critical inquiry into all three is essential."[6]

---

[6]**Dalai Lama, 34 – 5.**

The Dalai Lama's second visit to Europe was in 1979. Keep in mind in what follows that this was four years before his visit to CERN. It was during his 1979 visit that the Dalai Lama began his studies of quantum theory. In a way that he does not explain in his book, he acquired two "tutors", Carl Friedrich von Weizsäcker and David Bohm. Two more disparate figures it is impossible to imagine. Von Weizsäcker, who was born in Kiel in 1912, came from an aristocratic German family. He had inherited the title of *Freiherr*—baron. His father, Ernst, was a diplomat and a Nazi. He became Secretary of State under the Foreign Minister, Joachim von Ribbentrop. He also became a high official in the SS. In 1947, he was tried and convicted at Nuremburg for having abetted the deportation of French Jews to Auschwitz. At his trial, his defense lawyer was his son, Richard, who later became the President of Germany. Weizsäcker was convicted and sentenced to seven years, five of which he served.

C.F. von Weizsäcker studied physics with various people, including Niels Bohr and Werner Heisenberg. Heisenberg gave him the problem of working out the detailed optics of what is called the "Heisenberg microscope." This is like any other microscope, except that Heisenberg used it in a "thought experiment" to determine the position and momentum of a particle. The idea is that a quantum of radiation—many quanta, in a realistic situation—is scattered off the particle and arrives at the microscope lens. The resolving power of the microscope—the capacity to determine the particle's position—is proportional to the wavelength of the quantum. But each quantum carries a momentum that is inversely proportional to the wavelength. Thus, the more precisely one measures the position with the microscope the more momentum one imparts to the particle. This is a textbook example of the Heisenberg uncertainty principle. But Heisenberg asked von Weizsäcker to work out the details in a realistic case. After taking his Ph.D. Weizsäcker became an assistant to Lise Meitner in Berlin.

While he was with Meitner he created what is known as the "Weizsäcker semi-empirical mass formula." This formula enables one to plot the masses of all the nuclei—except for the very lightest ones—in terms of a small number of empirically determined parameters. Meitner made use of this formula on the famous walk in the woods she made in December of 1938 with her nephew Otto Frisch. She was able to argue that the experiments of her former Berlin colleagues, Otto Hahn and Fritz Strassmann, showed that the uranium nucleus had fissioned. Weizsäcker also proposed a process by which the Sun generates its energy. He did not work out the details. This was done by Hans Bethe who had independently proposed the same process.

When, in the fall of 1939, German Army Ordinance drafted a number of scientists to work on nuclear energy, including the prospects for nuclear weapons, Weizsäcker and Heisenberg were among them. One of Weizsäcker's contributions in the summer of 1940 was a proposal to use what was later called plutonium in a nuclear weapon. In 1945, he was one of the German scientists who was captured and interned at Farm Hall, near Cambridge.

Weizsäcker, and Heisenberg for that matter, were "non-Nazis"—a term that was used at Farm Hall. Neither of them joined the Party. On the other hand Heisenberg certainly wanted, in the beginning, for the Germans to win the War, and both of them, despite what they said later, were quite willing to work on an atomic bomb for Hitler. When Strasbourg (Strassburg in German) fell to the Germans, the university was taken over and Aryanized. Weizsäcker had no problem accepting a professorship there in 1942. His tenure was not long as the city fell to the Allies in 1944. After the Germans were released from Farm Hall, Weizsäcker returned to Germany and played an important role in resurrecting German academic science. He was also an outspoken opponent of nuclear armament for Germany. His days as a creative scientist were over but he wrote a good deal about philosophy and in particular, about quantum mechanics. In quantum mechanics, he was a Bohr positivist. I will explain this more fully later, but in essence he agreed with Bohr that there was no hidden classical substratum underlying the quantum theory. The theory with all its probabilities and uncertainties had to be accepted for what it was. Weizsäcker died on April 28, 2007.

David Bohm was the absolute antipode to Weizsäcker, which makes the Dalai Lama's choice of these two men as his tutors so remarkable. Bohm was born in Wilkes-Barre, Pennsylvania in 1917. His mother and father were Jewish immigrants from Eastern Europe. His father owned a furniture store and was the assistant to the local rabbi. After graduating from Penn State, Bohm went to California where Robert Oppenheimer became his thesis advisor. Bohm became very involved with left-wing political organizations, including the Young Communist League. When the War came, Bohm could not get clearance to go to Los Alamos with Oppenheimer. Considering the people who were cleared, including Oppenheimer's brother who had been a member of the Communist Party, this is rather odd. Bohm remained in Berkeley where he did work connected to the atomic bomb project. After the War, Bohm joined the faculty of Princeton University. In the early 1950s two things happened to Bohm which changed his life. On the one hand he had been called to testify before the House Un-American Activities Committee where he had taken the Fifth Amendment. Princeton suspended him and Oppenheimer suggested that he leave the country—rather strange advice considering Oppenheimer's own situation. Bohm took this advice, going first to Brazil, then to Israel, and finally to England, where he died in 1992.

The second thing that happened was that in 1951, Bohm published his text, *Quantum Theory*.[7] This was the only text at the time that I knew of where the conceptual foundations of the theory were discussed at length. People who knew Bohm then have told me that arriving at the position he took in the text had required some personal struggle. For some of the Marxist–Leninist dialecticians, the Bohr interpretation was anathema. An electron, to take an example, is neither a particle nor a wave. Depending on the experiment it can manifest itself as one or the other, or both. To Bohr these were "complementary" descriptions that reflected the limitations of language. Our language is classical and we do our best to apply it to quantum mechanics. But this, in some hands, can morph into the notion that we create reality—the world is a will and an idea. It is this idealism that this school of Soviet dialecticians objected to. It is what apparently troubled Bohm when he first learned the theory. But his textbook reflects none of this. It is orthodox Bohr.

Bohm was quite surprised when Einstein unexpectedly called him. He had read the book and told Bohm that it was the best attempt at refuting his views that he had ever seen. Einstein believed that the quantum mechanical description was not complete, while Bohm argued in his book that it was. Einstein accepted the consequences of the uncertainty principle but thought that this and the other aspects of the theory would emerge as limits of some sort of classical theory which he never succeeded in creating. He invited Bohm to come and visit him to discuss these matters. Exactly what happened in this discussion I have never been able to find out. I once wrote Bohm to ask him. He answered my letter but not the question. It seems that Bohm was persuaded that the orthodox interpretation of the theory—the one that he had so brilliantly presented in his book—left something to be desired. It also seems as if Einstein proposed something fairly specific because he later expressed his disappointment that Bohm had not carried it out.[8] What Bohm

[7]**David Bohm, *Quantum Theory,* Prentice Hall, New York.**

[8]**I am grateful to Bruce Rosenblum for telling me that on a visit to Princeton he heard Einstein say that "David" did not do what he had asked him to do. Einstein did not explain what he had asked Bohm to do.**

did do was to produce a theory of "hidden variables" which was completely deterministic and classical but which reproduced all the results of the quantum theory so long as effects of the theory of relativity could be neglected.

The Dalai Lama, judging from his book, was very fond of Bohm. Like Popper, and himself, he also saw in Bohm someone who had been exiled from his own country. The Dalai Lama writes,

> "I particularly admired Bohm's extraordinary openness to all areas of human experience, not only in the material world of his professional discipline but in all aspects of subjectivity including the question of consciousness. In our conversations, I felt the presence of a great scientific mind which was prepared to acknowledge the value of observations and insights from other modes of knowledge than the objective scientific."[9]

Later in the book he writes about Bohm's notion of "implicate order", which is to replace reductionism. This is a kind of holistic "flow"—perhaps comparable to the Buddhist concept of "wholeness."

The Dalai Lama remarks, "I notice that there is a group of scientists and philosophers who appear to believe that scientific thinking derived from quantum physics could provide an explanation of consciousness. I remember having some conversations with David Bohm about his idea of an 'implicate order', in which both matter and consciousness manifest according to the same principles. Because of their shared nature, he contends, it is not surprising that we find a great similarity of order between thought and matter. Although I never fully understood Bohm's theory of consciousness, his emphasis on a holistic understanding of reality—including both mind and matter—suggests an avenue by which to look for a comprehensive understanding of the world."[10]

In his book the Dalai Lama presents his understanding of the problems in the interpretation of quantum theory. It comes in a long (and I think, very remarkable) statement which I will quote in full and then comment on.

---

[9] **Dalai Lama, 30.**
[10] **Dalai Lama, 128.**

He writes, "In physics, the deeply interdependent nature of reality has been brought into focus by the so-called EPR paradox—named after its creators, Albert Einstein, Boris Podolsky and Nathan Rosen—which was originally formulated to challenge quantum mechanics. Say a pair of particles is created[11] and then separates, moving away from each other in opposite directions—perhaps to greatly distant locations, for example, Dharamsala, where I live, and, say, New York. One of the properties of this pair of particles is that their spins must be in opposite directions—so that one as "up" and the other will be found to be "down." According to quantum mechanics, the correlation of measurements (for example when one is up, then the other is down) must exist even though the individual attributes are not determined until the experimenters measure one of the particles, let us say, in New York. At that point, the one in New York will acquire a value—let us say, up—in which case the other particle must simultaneously become down. These determinations of up and down are instantaneous, even for the particle in Dharamsala, which has not itself been measured. Despite their separation, the two particles appear as an entangled entity. There seems, according to quantum mechanics, to be a startling and profound interconnectedness at the heart of physics."

The Dalai Lama goes on, "Once at a public talk in Germany, I drew attention to the growing trend among serious scientists of taking the insights of the world's contemplative traditions into account. I spoke about the meeting ground between my own Buddhist tradition and modern science—especially in the Buddhist arguments for the relativity of time and for rejecting any notion of essentialism. Then I noticed von Weizsäcker in the audience, and when I described my debt to him for what little understanding of quantum mechanics I possess, he graciously commented that if his own teacher Werner Heisenberg had been present, he would have been excited to hear of the clear, resonant parallels between Buddhist philosophy and his scientific thoughts."

The Dalai Lama concludes, "Another significant set of issues in quantum mechanics concerns the question of measurement. I gather that, in fact, there is an entire area of research dedicated to this matter. Many scientists say that the act of measurement causes a "collapse" of either the wave or the particle function depending on the system of measurement used in the

[11]**The Dalai Lama, *The Universe in a Single Atom*, Morgan Road Books, New York, 2005, 2 – 3. This statement of the Dalai Lama does not make it clear that these particles have to be created in what is called an "entangled" state. This is explained in chapter 6. Dalai Lama, 64 – 6.**

experiment; only upon measurement does the potential become actual. Yet we live in a world of everyday objects. So the question is, how, from the point of view of physics, do we reconcile our commonsense notions of an everyday world of objects on the one hand and the bizarre world of quantum mechanics on the other? Can these two perspectives be reconciled at all? Are we condemned to live with what is apparently a schizophrenic view of the world?"[12]

As I was reading this I could not help thinking of John Bell. It was his analysis of the Bohm version of the EPR experiment, which is what the Dalai Lama is discussing, which showed that hidden variable versions of quantum theory would inevitably involve transmission of influences at speeds greater than that of light, making them, on their face, incompatible with relativity. The questions about how the quantum and the classical worlds co-exist were the ones that bothered Bell the most. None of the proposed solutions entirely satisfied him. In Bohm's model the world is classical, but non-local. In other models the world is entirely quantum mechanical which requires explaining where everyday classical physics fits in. Some people believe that the world is divided into patches, some of which are quantum mechanical and some classical, which leaves the question of characterizing the patches. If only Bell and the Dalai Lama had gone off together to discuss these things by themselves. Just imagine what they could have taught each other.

---

[12]**Dalai Lama, 55 – 6. Dalai Lama, 64 – 6.**

Chapter 3

Léon Rosenfeld

"Dear($\sqrt{\text{Trotsky}\times\text{Bohr}} = \text{Rosenfeld}$)"

Wolfgang Pauli[1]

"Is it not delightful that poor Bohr's only supporters in Paris should be this logical couple [de Broglie and Destouches], while all the youth are up in arms against him "under the banner of Marxism"? Poor Marx too, I would add, since I belong, as you know, to the almost extinct species of genuine Marxists: the kind of theology dished up under this name today is just as repulsive to me as to you, perhaps even more so because I see it against a background of what Marx really meant."

Léon Rosenfeld[2]

Niels Bohr had great difficulty in the actual writing of a paper, or in giving a lecture for that matter. One would be tempted to say that it was some curious form of dyslexia. This was especially true in his mature years as his output became more and more philosophical—less mathematical. To deal with this, Bohr hired assistants, part of whose job was to act as sounding boards and amanuenses. When none of these assistants were available, he pressed anyone who seemed both qualified and at hand into

[1]This and much of the biographical material on Rosenfeld I have taken from Anja Skaar Jacobsen, "Léon Rosenfeld's Marxist Defense of Complementarity", Historical Studies in the Physical and Biological Sciences, Vol.37, Supplement. 2007 3-37. I will refer to this as AS. This quote is from AS, 4. There does not seem to be a full biography of Rosenfeld. Pauli's joke was in part mathematical. He was saying that Rosenfeld was the geometric mean between Bohr and Trotsky.

[2]AS, 24.

this duty. One of these people was the late Abraham Pais. He wrote about this experience in his biography of Bohr, *Niels Bohr's Times.*[3] Pais writes,

> "Bohr devoted tremendous effort and care to the composition of his articles. However, to perform the physical act of writing, pen or chalk in hand, was almost alien to him. He preferred to dictate. On one of the few occasions that I actually did see him write himself, Bohr performed the most remarkable act of calligraphy I shall ever witness." " ... as we were discussing the address that Bohr was to give on the occasion of the tercentenary celebration of Newton's birth, Bohr stood in front of the blackboard (wherever he dwelt, a blackboard was never far) and wrote down some general themes to be discussed. One of them had to do with the harmony of something or other. So Bohr wrote down the word 'harmony'."

In his book Pais gives what he recalls as Bohr's written version of the word 'harmony'. It is a squiggle that could be anything. Pais goes on,

> "However, as the discussion progressed, Bohr became dissatisfied with the use of 'harmony'. He walked around restlessly. Then he stopped, and his face lit up. 'Now I've got it. We must change harmony to uniformity ... ".[4]

Bohr then picked up the chalk and added to the previous squiggle a single dot, and then he put down the chalk with a look of great satisfaction.

As I can attest to, Bohr as a lecturer was something of a disaster. Not only was he often inaudible but he would do things like the one that Pais also describes. In the course of a lecture Bohr made a point and then changed his mind. He said "but" and was silent until whatever followed the "but" played out in his mind. Then he resumed the lecture as if he had spoken his thought, never actually doing so. Over the years Bohr had a number of assistants and collaborators. One of the most interesting and long-lasting was the Belgian physicist Léon Rosenfeld. They collaborated for over thirty years from 1930, until Bohr's death in 1962.

Rosenfeld was born on the 14th of August 1904 in Charleroi, Belgium, making him a couple of years younger than Pauli or Heisenberg. His father

[3] **Abraham, Pais, *Niels Bohr's Times,* Oxford University Press, New York, 1991.**

[4] **Pais, 9 – 10.**

was an electrical engineer who died in a tragic accident when Rosenfeld was fourteen. After high school he entered the University of Liège, graduating with high distinction in mathematics and physics in 1926. He then won a fellowship that enabled him to study in Paris at the École Normale Supérieure. Among his teachers was the physicist Paul Langevin, who had strong left-wing views and who finally joined the Communist Party in 1944. He was one of the Parisian scientists in the 1920s who were actively engaged with social problems. Rosenfeld took part and it was the beginning of his lifelong preoccupation with Marxism. As far as I can tell, Rosenfeld never joined the Party. He did visit Russia in 1934 with Bohr. His Marxism pretty much stopped with Marx and Engels. Lenin had already strayed from the True Belief and Stalin was beyond the pale.

I will have much more to say about Rosenfeld, Marxism and the struggle for the soul of quantum theory later, but here let me say a few words about Lenin's book *Materialism and Empirio-criticism*,[5] which he wrote while in London and Geneva, on exile from Russia, in 1908. While Rosenfeld does not refer to this book in any of his writings that I have seen, it became the basis of the Soviet Communist ideological canon when it came to science. Its echoes can be seen in the debate, often vitriolic, that began to take place in the 1940s over the interpretation of quantum theory.

In some ways, reading Lenin's book is like watching one of those old black and white films in which comic actors throw pies at each other. Opponents "cravenly hide" or talk "gibberish". Typical sentences read, "The Russian Machists will soon be like fashion-lovers who are moved to ecstasy over a hat which has already been discarded by the bourgeois philosophers of Europe."[6] Machists are variously "muddled", "brainless" and show "appalling ignorance." All good clean fun until it became possible in the 1930s to send someone to the *Gulag,* or worse, who disagreed. Why Mach?

Ernst Mach was an Austrian physicist and philosopher of science who was prominent at the end of the 19th century and the beginning of the 20th. His book, *The Science of Mechanics,* played an influential role in the formation of ideas of the young Einstein. Einstein was especially struck by Mach's polemic attack on Newton's idea of an absolute space and time beyond the reach of ordinary clocks and rulers. Mach believed that scientific theories were simply the economical description of observations. He no doubt subscribed to the motto of the 17th and 18th century divine, Bishop George Berkeley, who famously stated "*esse est pecipi*"—to be is to be

[5]**V.I.Lenin, *Materialism and Empirio-criticism,* Progress Publishers, Moscow, 1987.**
[6]**Lenin, 79.**

perceived. Lenin states in his book several times that all Mach did was to plagiarize Berkeley. Mach was an anti-atomist. He once had an exchange with his University of Vienna colleague Ludwig Boltzmann, a very great physicist to whom we owe, among other things, our notion of entropy in terms of probability. Mach said that he did not believe in atoms. When Boltzmann objected, Mach asked, "Have you seen one?" For Lenin, the emphasis would be on the "you"—Have YOU seen one?" Your personal observation would be necessary before you could say that atoms exist. For Lenin, this was the slippery slope at the bottom of which is the notion that there is no objective world outside yourself. Here, as we shall see, are the seeds of the objection of some Soviet scientists and philosophers about Bohr's interpretation of the quantum theory—to be is to be measured. Interestingly, Einstein is never mentioned in Lenin's book. However, my teacher, the noted philosopher of science Philipp Frank, gets a dig. Lenin writes, "The French mathematician Henri Poincaré", we read in the work of the Kantian Philipp Frank, "holds the point of view that many of the most general laws of theoretical natural science (e.g., the law of inertia, the law of conservation of energy, etc...) of which it is often difficult to say whether they are empirical or of *a priori* origin are, in fact, neither one nor the other, but are purely conventional propositions depending on human discretion...".[7] Professor Frank used to cite this with a good deal of amusement, especially at the notion that he, one of the founders of the Vienna Circle, the group of logical positivists, was a "Kantian". Now back to Rosenfeld.

After his Paris sojourn Rosenfeld became an assistant to Max Born in Göttingen during the period 1927–29. This was just after quantum mechanics had been created, in which Born played an essential role. I would imagine that this must have been an incredibly interesting time for a young physicist and a little discouraging too, when one saw that people such as Heisenberg, Pauli and Paul Dirac, who were about the same age, were making such profound contributions. I think that it is fair to say that Rosenfeld never did physics at this level. Probably the most important physics paper that Rosenfeld ever wrote was the one he published with Bohr in 1933. He had gone to Copenhagen in 1930. This paper had to do with whether, in the light of quantum mechanics and its uncertainties, the field strengths of electric and magnetic fields can be measured. They showed that it was not possible to measure these fields, unlike the classical

[7] **Lenin, 48.**

case, at a single point, but only over a small region of space. If one does this carefully the Heisenberg uncertainty relations are respected. Rosenfeld must have impressed Bohr on a personal level because he soon became Bohr's sounding board.

We get a glimpse of how this must have been from Rosenfeld's account of Bohr's reaction to the 1935 paper written by Einstein, Boris Podolsky and Nathan Rosen. By this time Einstein had settled in Princeton and Podolsky and Rosen were working with him there. It was written in English largely by Podolsky and titled, "Can Quantum-Mechanical Description of Physical Reality be Considered Complete?" Their answer was "no", and they illustrated this with a "thought experiment" involving the measurement of the positions and momenta of widely separated particles. Here is not the place to go into this except as an illustration of the Bohr–Rosenfeld collaboration. This is nicely described in Pais' biography of Bohr. Here is Rosenfeld's recollection as Pais quotes,

> "This onslaught came down upon us as a bolt from the blue...As soon as Bohr heard my report of Einstein's argument everything else was abandoned, we had to clear up such a misunderstanding at once". "Rosenfeld has told me that Bohr was infuriated when he first was informed. The next day he appeared at his office all smiles, turned to Rosenfeld and said, 'Podolski, Opodolski, Iopodolski, Siopodolski, Asiopodolski, Rasiopodolski'. Poor Rosenfeld was bewildered. Bohr explained he was quoting from a Holberg play in which a servant suddenly starts talking gibberish... When Rosenfeld remarked that Bohr had evidently calmed down, Bohr replied, 'That's a sign that we are beginning to understand the problem...They do it smartly but what counts is to do it right.' "[8]

Another instance occurred in the winter of 1939. After his Christmas vacation in Sweden with his aunt Lise Meitner, Otto Frisch returned to Copenhagen where he was then working. He immediately told Bohr about their analysis of Otto Hahn and Fritz Strassman's experiments on uranium. Meitner and Frisch realized, which Hahn and Strassmann had not, that fission had been discovered. Bohr's reaction was the same as that of every physicist I have spoken to, or read about, who first learned about fission

[8]**Pais, 430– 31.**

then. How could we possibly have been so stupid as not to have predicted it? The theory was there in front of their eyes. Bohr was just leaving for the United States to spend the spring in Princeton. He was taking Rosenfeld. Bohr had a cabin in which a blackboard had been set up. He spent much of the voyage going over all the arguments for fission. He had made an arrangement with Frisch to say nothing until he and Meitner published their paper. This took the pressure off of them to hurry. But Bohr forgot to tell Rosenfeld. As soon as they landed, Rosenfeld went off to Princeton and began to spread the word that very night. By the next day, the news had propagated all over the country and experiments were underway. The race for the atomic bomb was on.

In 1939, the Dutch physicist George Uhlenbeck emigrated to the United States. He had been a professor at the University of Utrecht. This left a vacancy at the University and Rosenfeld was invited to fill it. Until that time he had been commuting between the University of Liège, where he had a position, and Copenhagen, where he went to work with Bohr. Uhlenbeck left behind some unfinished business, including Pais. Pais had been studying at the University of Amsterdam and had decided that the physics being taught and practiced there was not up to date. He began to visit Uhlenbeck in Utrecht and by 1938, he had enrolled at the university. When Uhlenbeck left, he introduced Pais to the very distinguished Dutch physicist Hendrik Kramers who was at Leiden, who became Pais' mentor. But Rosenfeld became Pais' official thesis advisor in 1940, when he took up his new post. In the meanwhile, the Germans had invaded Holland. Jews were banned from all civil service positions, including teaching at universities. Pais lost his assistantship. Rosenfeld arranged for him secretly to share the assistantship of his non-Jewish successor. In 1941, the Germans decreed that no Jew could be awarded a doctorate after June 14th. Pais obtained his on June 9th, the last Jew to be awarded such a degree until after the war.

Pais felt at first safe because of his former university status. But in 1943, the Germans began rounding up Jews from the universities, intending to send them East for extermination. Pais went into hiding. His sister and brother-in-law did not and were captured, sent to concentration camps, and killed. Few people knew about Pais' hiding places. One of them was at the Kramers', who visited him and discussed physics. Pais' last hiding place was an apartment which he shared with his friend, Lion Nordheim. They were betrayed and arrested by the *Gestapo*. Pais told me that he thought he was saved by his knowledge of German. He managed to persuade his captors that he was an innocent scholar. Nordheim was not so fortunate. He shared a cell with Pais and Pais told me how he tried to comfort Nordheim the night before his execution. Very soon thereafter, Holland fell to the Allies and Pais was liberated. Rosenfeld was Jewish on his father's side.[9] Also, he had married a non-Jew. This, and the fact that he was a Belgian citizen, enabled him to remain in his post at the university until 1944. He then decided it was too dangerous and went into hiding in secret quarters in his attic. Although the Germans made a search he remained undiscovered until Holland was liberated in 1945. Rosenfeld stayed in Holland until 1947, when he went to the University of Manchester to direct its physics department. He stayed there until 1958, when he became the first professor at the then newly-created Nordic Institute for Theoretical Atomic Physics (NORDITA) in Copenhagen. There he remained until his death in 1974.

Rosenfeld's wartime experiences seemed to have hardened his character. He is often described as a reserved, gentle and even timid man. But after the War, when he began attacking people who were making critical attacks on Bohr's interpretation of the quantum theory, he was ferocious. His letters and articles are scathing. To understand the issues we must backtrack some years.

The colloquy between Bohr and Einstein about quantum theory began in the late 1920s. Einstein presented various thought experiments which purported to show that the Heisenberg uncertainty principles led to contradictions. Bohr shot them down. This persuaded Einstein that there were no contradictions related to the uncertainty principles. On the other hand, he felt that the theory, despite its successes, did not give a complete description of reality. This was the purport of his paper with Podolsky and Rosen. But there is even a much simpler example than the one they gave in their paper. If one has a sample of radioactive elements, the theory does not tell you which of the atoms will next decay. It only gives you the probability that one might decay. Moreover, there is no way, using the quantum theory that you learn in school, to even formulate the proposition that a particular atom decayed yesterday at four o'clock. The quantum theory as usually presented deals with the probability of future events and not a description of the past. If you think these issues deserve an explanation then you might think that Einstein had a case.

While this clash of Titans was going on most physicists were largely indifferent. There was too much to do. The Nobelist I.I. Rabi once said

[9] **I am grateful to Anja Skaar Jacobsen for this information.**

to me that he thought Bohr was very profound about things that didn't matter. Rabi was trying to design experiments that would show a breakdown of the theory. He never found any. After a visit to Princeton in 1935, Robert Oppenheimer wrote to his brother, Frank, "Princeton is a madhouse: its solipsistic luminaries shining in separate and helpless desolation. Einstein is completely cuckoo."[10] Of the founders, Heisenberg occasionally made philosophical comments: his version of Bohr's ideas. Erwin Schrödinger, who was a little older than the others and very philosophically minded, sided with Einstein. He had been skeptical of the theory for some time. In 1935, he published a paper in the German journal *Naturwissenschaften* whose translated title is "The Present Situation in Quantum Mechanics." In it he introduced two things that have endured. One was a term of art—"entanglement." This describes the quantum state that is prepared, for example, in the Einstein–Podolsky–Rosen experiment. Even when the particles are widely separated in space—out of ordinary communication with each other—their properties are correlated. If you select one of these properties, quantum theory predicts the probability, upon a measurement on one of these particles, of finding some possible value of this property. Once having found this, there is a correlation to the probable outcomes of measurements made on the distant particle. In this sense the particles are "entangled", despite the fact that no force is involved. The second thing Schrödinger introduced in that paper which has endured is the "cat."

In this thought experiment, a cat is confined inside a sealed container. The container contains a radioactive sample. When it disintegrates, it produces ionization which unleashes a vial of poisonous gas, enough to kill the cat. At any given time there is some probability of such a radioactive decay. In the meanwhile, if one wants to make a quantum mechanical description of the state of the cat, and wants to be provocative about it, one can say that at this time the cat is both alive and dead. A Bohrian positivist would probably say "prove it." You would then open the container to find either a live or dead cat. I paid a visit to Schrödinger in his apartment in Vienna not long before his death in 1961. There were no cats. I was told that he did not like cats. Another founder, Dirac, as far as I know, never made any published comments about cats. He was too busy applying the theory. For example, the union of relativity and quantum mechanics that

[10]**Eds Alice Kimball Smith and Charles Wiener, *Robert Oppenheimer*; *Letters and Reflections*, Harvard University Press, Cambridge, 1980, 190.**

he created had, as an offspring, anti-matter. About this time he noted, with his usual pith, that quantum mechanics explained most of physics and all of chemistry.

Bohr, however, soldiered on during the 1930s. He gave lectures and wrote papers. Out of this emerged something that is often referred to, though never by Bohr, as the "Copenhagen interpretation" of the quantum theory—the *Copenhagener geist*. Since this was never spelled out formally, there are some differences of opinion as to what precisely this is. However, all are agreed that Bohr's notion of "complementarity" plays an essential role. When I think of complementarity, what Saint Augustine said about time comes to mind. "What then is time? If no one asks me, I know what it is. If I wish to explain it to him who asks, I do not know." Substitute "complementarity" for "time." Bohr applied this idea, of which he was very proud, to practically all forms of human activity. I will confine myself to physics and give two examples, one classical and one quantum mechanical. I begin with the classical.

To this end I consider a container which holds molecules of some gas. I then heat the walls of the container. This agitates the atoms out of which the walls are composed. When molecules of the gas collide with the walls, they are in turn agitated, and this agitation communicates itself to the entire gas. From a macroscopic point of view we say that the gas has been "heated", and if we insert a thermometer it will show a rise in temperature. Temperature is a macroscopic property of the gas. If instead, I choose to measure the energy of one of the molecules, I cannot in this measurement also measure the temperature of the gas, which is a collective property of the ensemble of molecules and requires a different sort of apparatus to measure. Thus, these two kinds of measurements are, in Bohr's sense, "complementary." They apply to two mutually exclusive descriptions, both of which are valid in their domains of applicability. We must make a choice as to which one we wish to apply. Now to the quantum mechanical example.

In 1922, the French nobleman Louis de Broglie, who was working towards his *doctorat*, went to his thesis advisor Paul Langevin with an idea. Langevin found it so radical that he sent it to Einstein to seek his advice. Einstein said that he found it interesting and had been thinking along somewhat similar lines, although he had not published anything. De Broglie wrote a brief note which he published in the *Comptes rendus*.[11] Reading this note is an odd experience. One knows where it led, but from the

[11] *Comptes rendus* **Vol 177, 1923, 507–10.**

present point of view it seems pretty confused. In more modern language, what de Broglie was suggesting, however obscurely, is that electrons—he deals specifically with electrons—which had until that time been assumed to be corpuscular—had a wave nature as well. He calls them "fictitious waves." He never really says what we would say now, namely that the wavelength of these waves is inversely proportional to their momenta. Nor does he present any suggestion for an experimental test of this notion. Einstein did. Einstein said that if this was true, and if electrons were sent through a diffraction grating of a suitable size, diffraction patterns would emerge. This was the way light was shown to be wave-like at the beginning of the 19th century. In 1927, experiments by Clinton Davisson and Lester Germer, and independently by George Thomson, were performed on electrons in which the diffraction patterns were observed. The "duality" of electrons was established. This was the counterpart to the duality of light-particles and waves that Einstein had proposed in 1905.

The two aspects—particle and wave—were complementary in Bohr's sense. This manifests itself in the expression for the wavelength in terms of the momentum. "Wavelength" is evidently a wave phenomenon and momentum is something one associates with particles. Suppose we do an experiment with a diffraction grating allowing the electrons to pass, one at a time. If we collect the electrons on a photographic plate, the regions of maximum and minimum intensity will spell out the diffraction pattern, but each electron leaves a localized spot on the plate, something that a wave does not do. Which aspect is revealed depends on our choice of experiment—complementarity. If one is not careful with language—and at times Bohr wasn't—this slips into idealism. Indeed, some writers such as Heisenberg, did think that it was idealism—a "participatory universe." In the most extreme cases, the claim is that it is our presence that brings the laws of physics about. It was just here that the conflicts with some of the Soviet scientists took place. Vladimir Fock, who died in 1974, was a first-rate Russian theoretical physicist. He was not quite in the Nobel category, but some of the things he created are still taught. In 1935, he translated into Russian the Einstein, Podolsky and Rosen paper. He agreed with Bohr's criticisms. Moreover he saw "complementarity" as a manifestation of Engels' second dialectical law:[12]

"The law of unity and the conflict of opposites."

[12]**The first and third are respectively the transformation of quantity into quality and the negation of the negation.**

Fock also argued that Einstein's theory of relativity was a manifestation of "materialism". How sincere he was I do not know, but so long as Bohr did not stray into language which seemed to imply that matter did not have objective properties, Fock was all right. He seems to have come through the Great Purges of 1937 and 1938 unscathed. Others were not so fortunate.

Matvel Petrovich Bronstein was a physicist and astrophysicist of great promise. In the winter of 1938 he was put on a list of tens of thousands who were to be arrested, charged and then executed. "Innocent until proven guilty" was considered a bourgeoisie notion. On February 18, 1938, he was charged and executed on the same day. What his crime was, was never explained. His wife was told that he had been sentenced to ten years in labor camps without the right of correspondence. He was thirty-two.

Lev Davidovich Landau was, in most people's opinion, the greatest Russian theoretical physicist of the 20th century. He was awarded the Nobel Prize in 1962. In 1938, he was arrested on a trumped up charge and held in an NKVD prison. If a very courageous colleague, Pytor Kapitsa, had not written a letter directly to Stalin, Landau might never have gotten out. Unlike Fock, both of these men were Jewish, which made everything worse for them. In June of 1941, the Germans invaded the Soviet Union and the compatibility of dialectical materialism and complementarity was not high on the list of priorities.

I began my graduate studies in the early 1950s. English translations of some excellent Russian texts were available. I remember in particular a book entitled *Quantum Electrodynamics,* written by A.I. Akhiezer and V.B. Berestetski, whose translation appeared in 1953. You would be reading along in the technical material when suddenly, and with no warning, there would be a paragraph informing you how this work was connected to dialectical materialism. Then the physics went on as if nothing had happened. I always thought of these as little commercial messages from the "sponsor." It did not occur to me to inquire what circumstances must have led to this. In fact, after the War, the ideological struggle for the soul of quantum theory in the Soviet Union took a rather dramatic turn. In the 1930s the views of Fock were criticized but these criticisms could be largely ignored. At least, Fock more or less ignored them. But by the late 1940s things had changed. Fock now had critics he could not ignore. Among them was the physicist D.L. Blokhintsev.

Blokhintsev, who was a few years younger than Fock, was a theoretical physicist of considerable distinction. He was, one gathers, a bit higher in the ranks of the establishment. He ended his career as Chairman of the department of theoretical nuclear physics at the Moscow State University. In 1944, he published the first edition of a text on quantum theory. It went through several editions and translations. In the first edition he embraced at least part of the Copenhagen interpretation.[13] To understand this, and Blokhintsev's later changes, we have to back up a little.

De Broglie was very vague about the meaning of his waves. He did not provide an equation that described their propagation in space and time. This was done in 1926 in a series of four magisterial papers by Erwin Schrödinger. The Schrödinger wave equation tells us how the "wave function", usually denoted by ψ, evolves in space and time. But what does it mean? At first Schrödinger thought that it was a guide wave that guided the particles. This guide wave was to have the same physical properties as any wave. But quite soon it was shown that this interpretation was untenable. In 1927, the Göttingen theoretical physicist Max Born proposed that these waves were waves of probability. A particle was likely to be found where the amplitude of the wave was large. This is the interpretation we still have.

The question which then arose is whether this probability is all there is? To this, Bohr answered "yes." On this, Fock agreed. In his introduction to his translation of the Einstein–Podolsky–Rosen experiment paper, he wrote,

"In quantum mechanics the conception of state is merged with the conception of 'information about the state obtained as a result of a specifically maximally accurate operation.' [a measurement of, say, the position of a particle]. In quantum mechanics the wave function describes not the state in the usual sense, but rather 'information about the state.' "[14]

We are in the position of someone at a horse race who is allowed to know only the odds. No other information, such as the breeding of the horses, is available. If we go further and say that no such information is ever available to anyone we get some sense of the quantum mechanical situation.

In the 1944 edition of his book, Blokhintsev did not seem to disagree with this, but in the 1949 edition, he wrote, "The chapter which concerns concepts of the state in quantum mechanics has been changed, and the clarity of the discussion of the uncertainty relationships has been improved. In the new edition of the book,

---

[13]**For a very enlightening discussion of these matters see Loren R. Graham," Quantum Mechanics and Dialectical Materialism", Slavic Review, No.3, September, 1966, 381–410 and the response from Fock, 411– 413.**
[14]**Graham, 384.**

ideological questions connected with quantum mechanics are also considered, and the idealistic conceptions of quantum mechanics which are now widespread abroad are subjected to criticisms."[15]

In the interval between the editions there had been some strong and potentially dangerous criticisms by various Soviet authors of the Copenhagen interpretation. In the 1949 edition Blokhintsev speaks of the wave function controlling the destiny of what he called "microparticles" and these had an objective reality. He and Fock had some spirited exchanges which did not seem to change either of their minds. In 1966, Fock summarized his position by stating that

> "Dialectics plays an essential role in obtaining new outlooks on the external world and the appropriate ways of its description. To solve the contradictions between classical and quantum mechanical description, between causal and probabilistic laws, between wave properties and corpuscular properties of matter, between elementarity and mutual transformability of particles, between objectivity of micro-objects and the necessity of introducing measuring instruments to describe their properties, etc, etc—to solve any of these problems a dialectical approach is quite indispensable. Before proceeding to the solution one must first of all realize that the contradictions are of a dialectical nature so that they must vanish if a more general and more appropriate formulation of laws of nature is found. (In the early history of quantum mechanics an instance of that proceeding is given by the conscious application of dialectics by Bohr, who formulated the "semi-classical" quantum mechanical laws in spite of the apparent contradictions later solved by the [wave] quantum mechanics ... Dialectical materialism is a living and not a dogmatic philosophy. It helps to give the experience obtained in one of the domains of science a formulation of such generality that it may be applied to other domains."[16]

In contrast, in the 1964 edition of Blokhinstev, which I have, there is not a single mention of dialectical materialism.

---

[15]**Graham, 392.**
[16]**Fock, 412.**

Rosenfeld observed all of this with a good deal of contempt. As far as he was concerned, quantum theory did not have "interpretations." There is no quantum theory but the quantum theory and Bohr was its messenger. In 1972, the American physicist Henry Stapp published an article entitled "The Copenhagen Interpretation."[17] He sent a draft to Rosenfeld for vetting and got back the response

"I notice from your further letters with new title pages that you hesitate about the best title for your essay. I have no very strong views about this, but I would incline to prefer your March 31 title ["Quantum Theory, Pragmatism and the Nature of Space-time"], the reason being that it does not contain the phrase "Copenhagen interpretation," which we in Copenhagen do not like at all. Indeed, this expression was invented, and is used by people wishing to suggest that there may be other interpretations of the Schrödinger equation, namely their own muddled ones. Moreover, as you yourself point out, the same people apply this designation to the wildest misrepresentations of the situation. Perhaps a way out of this difficulty would be for you to say, after having pointed out what the difficulty is, that you make use of the phrase "Copenhagen interpretation" in the uniquely defined sense in which it is understood by all physicists who make a correct use of quantum mechanics ... ".[18]

Rosenfeld was at his most vitriolic when he addressed fellow Marxists. Typical of the genre is a letter he sent to Frédéric Joliot-Curie in 1952. Joliot, who had shared the 1935 Nobel Prize in Chemistry with his wife Iréne, had fought in the resistance, and was a member of the French Communist Party. Rosenfeld writes,

> "I think it is my duty to inform you of a situation which I consider quite serious. It concerns your "foals" Vigier, Schatzman, Vassails and the whole lot, all young intelligent people, and all full of desire to do well. Unfortunately, at the moment they are quite sick. They have gotten it into their heads that it is necessary persistently to shoot down complementarity and save determinism. The ill-fortune is that they have not understood the problem and—what is even worse—they have made no serious effort to understand it. I have done what I could to redeem them ... I have taken pains to do an explicit Marxist analysis of

[17]**Henry Stapp, American Journal of Physics, 20, 1072, 1098–1116.**
[18]**Stapp, 1115.**

> the question and clearly show the simultaneously dialectical and materialistic character of complementarity. As the only response, Schatzman sent me a polemical writing full of incorrect physics and quotations from Stalin, which he maintains, through a casuistry that frightens me, oppose physical evidence. This reveals a profound crisis among these young people and it is high time (if it is not already too late) to straighten them out. They are under a spell of a scholasticism which borrows the external forms of Marxism, but is as opposed to its genuine spirit as is the blackest Catholicism. The best Soviet physicists are subject to attacks from this scholasticism, which creates even more havoc in Moscow than in Paris. Surely it would be desirable that the French physicists show themselves capable of distinguishing the wheat from the chaff. At the moment, these young fanatics are the laughing stock of the theoreticians such as Destouches and de Broglie. I am ready to give you all possible support to redress the situation on the basis of the ideas explained in my paper. But it is up to you, my dear Joliot, to take the initiative."[19]

While the "foals" might be brought into line with a good talking-to, David Bohm was another matter. His apostasy was irredeemable. He was a *soi disant* Marxist who had abandoned the laws set down by Marx and Engels. Still worse, he had produced a version of quantum mechanics which, if adopted, would render Rosenfeld's whole life work irrelevant. In Bohmian mechanics there is no complementarity. There is no Einstein–Podolsky–Rosen paradox. Quantum mechanical measurements are just another set of ordinary interactions. The theory is deterministic and essentially classical. Classical particles are guided by waves determined by a Schrödinger equation. In the usual case, the solutions to the Schrödinger equation predict the probabilities of the outcomes of experiments. Here, the solutions provide the information that is needed to determine the trajectories of classical particles. For Rosenfeld, Bohm and his mechanics were beyond the pale.

Bohm's two articles in *the Physical Review* appeared in January of 1952. By this time Bohm was in Brazil. He had been there since the previous

[19]**Jacobsen, 25.**

November.[20] He had appeared before the House Un-American Activities Committee, taken the Fifth Amendment, been indicted by a grand jury for contempt of Congress and finally been acquitted on May 31, 1951. Bohm had joined the Communist Party in November of 1942, but only remained a member for some nine months. But he defended the Soviet regime until Stalin's death. After Bohm's appearance before the House committee, in which he took the Fifth Amendment, he received written notice that he was no longer to have access to the Princeton campus. How this was to have been enforced is unclear. In fact the department chairman Allen Shenstone told Bohm that he would be welcome in his office at any time and Bohm continued working with his graduate students.[21] Princeton University had put him on paid leave and then refused to renew his appointment in June of 1951. Some Brazilian physicists who were in Princeton were able to find him a job at the University of São Paulo. Not long after Bohm arrived in Brazil he was asked to present his passport to the American consul's office for "verification." It was confiscated and he was told that it would not be returned unless he returned to the United States to reclaim it. Fearing that he might be subject to further legal action if he returned, Bohm took Brazilian citizenship. It was some thirty years before his American citizenship was restored. By this time he had become an intellectual celebrity.

Bohm's three-year stay in Brazil was not an especially happy one. He found the country technologically backward and somewhat "dirty." He also managed to get into a fight with some of the physics professors. Bohm was given a stipend he could use to hire assistants. He chose them from the United States, which was deeply resented. Two of them, Ralph Schiller and George Yevick, happened to have been colleagues of mine at the Stevens Institute of Technology. In the three decades that I knew them, neither one ever said a word to me about this experience. In any event, there was a nasty fight over such appointments. At one point the German physicist Carl Friedric von Weizsäcker appeared in Brazil and seemed to side with

[20]For a very interesting account of Bohm in Brazil, see "Science and Exile: David Bohm, the hot times of the Cold War" by Olival Freire Jr., *Historical Studies on the Physical and Biological Sciences*, 36(1) 1–34, 2005. See also F. David Pate's biography of Bohm, *Infinite Potential*, Basic Books, New York, 1997. The reader should be aware that this book contains many mistakes of detail. For example, Pate claims that Bohm and Richard Feynman were graduate students of Oppenheimer together. Feynman was, in fact, a student of John Wheeler in Princeton. Allen Shenstone has become Allen Shelstone, Stirling A. Colgate becomes Stephen A. Colgate, etc.

[21]I am grateful to David Pines who was one of them for this information.

Bohm's opponents, whom Bohm referred to as "Nazis." It must have been with some relief all the way around when Bohm left for Israel, where he stayed for two years before finally settling in Great Britain at Birkbeck College in London. Here he remained until his death in 1992.

In the midst of Bohm's turbulent activities in Brazil came Rosenfeld to give lectures on the foundations of quantum theory which Bohm attended. Afterwards Bohm wrote a letter in which he said, "Prof. Rosenfeld visited Brazil recently, and we had a rather hot and extended discussion in São Paulo, following a seminar he gave on the foundations of the quantum theory. However, I think we both learned something from the seminar. Rosenfeld admitted to me afterwards that he could at least see that my point of view was a possible one, although he personally did not like it."[22]

Bohm must have read into Rosenfeld's reaction what he wanted to see. There was no way that Rosenfeld thought that Bohm's version of the quantum theory was a "possible" one. Sometime after his visit to Brazil, Rosenfeld wrote a letter to the British philosopher of science Lancelot Law Whyte, in which he said,

> "Why then ... do I bother to write all these articles and reviews? I am not concerned with Bohm's own salvation: he is past healing ... all I actually wish to do is to sound a warning to the bewildered students, who don't know what to make of it, and to interested outsiders, who—however clear-minded and critical—cannot reasonably be expected to be sufficiently familiar with the technicalities of the case to discover themselves the utter emptiness of the claims of Bohmism. I realize that I may give the impression of being "intolerant" and "dogmatic" in spite of the fact that I have been all these years advocating tolerance and free thinking ... and waging a fight against the modern brand of dogmatism which represents the greatest threat to our scientific tradition, viz pseudo-Marxist theology."[23]

At present this discussion of Marxism and science seems almost antique. One has the impression that Russian science has now moved on. I, at least, have not found the sort of ideological insertions in Russian papers that used to occur. But when I think of the past, I am reminded of an anecdote

[22]**Friere, 20.**
[23]**Jacobsen, 26.**

I heard. In 1944, Lavrenty Beria, the Chief of the Soviet secret police, was placed by Stalin to direct the nascent nuclear weapons program which involved some applied quantum mechanics. Beria discovered that certain physicists were straying off the ideological reservation and he complained to Stalin. Stalin allegedly said to him, "Leave my physicists alone. We can always shoot them later."

Chapter 4

A Double Slit

"We chose to examine a phenomenon which is impossible, absolutely impossible, to explain in any classical way, and which has in it the heart of quantum mechanics. In reality it contains the only mystery."

Richard Feynman[1]

"...The particle world is the dream world of the intelligence officer. An electron can be here or there at the same moment. You can choose. It can go from here to there without going in between; it can pass through two doors at the same time, or from one door to another by a path which is there for all to see, until someone looks, and then the act of looking has made it take a different path. Its movements cannot be anticipated because it has no reasons. It defeats surveillance because when you know what it is doing you can't be certain where it is, and when you know where it is you can't be certain what it's doing: Heisenberg's uncertainty principle; and this is not because you're not looking carefully enough, it is because there is no such thing as an electron with a definite position and definite momentum; you fix one, you lose the other, and it's all done without tricks, it's the real world, it is awake."

Tom Stoppard, Hapgood[2]

In the fall of 1947, I entered Harvard College as a seventeen-year-old freshman. I had no real ideas about my future, except that I was quite sure that it would have nothing to do with science. I had no interest in science. But at that time the President of the university was James Bryant Conant, a chemist who had played a very important role in the creation of nuclear weapons. This experience had left him with a good

[1]**R.P. Feynman, R.B. Leighton and M. Sands, *1963 Feynman Lectures,* vol 3, ch37.**

[2]**Tom Stoppard: *Plays*, Faber and Faber, London, 1999, 544.**

deal of concern about this new force in the world. One manifestation of this was the notion that every Harvard graduate should have some sort of scientific education. He knew, of course, that most of us would not have careers in science, but we might well have careers where we would have to make decisions that involved science. So under the rubric of "General Education" a set of natural science courses had been instituted which presented non-mathematical surveys of different scientific fields. Every Harvard graduate was required to pass one of them unless he—there were no 'she's—was a science major. You also had to be able to swim two laps of a twenty-five-yard long pool. Faced with the former obstacle, I consulted the *Harvard Confidential Guide to Courses*—a small booklet produced and sold by the undergraduate newspaper, the Harvard Crimson—to find out which course was considered the easiest. The vote among the undergraduates who took the various courses and compared notes was unanimous that it was Natural Sciences 3, taught by the late historian of science, I. Bernard Cohen.

In the interests of full disclosure I have to admit that I had had a physics course in high school. I have searched my memory to recall anything that I had learned in this course and have come up blank. It was my only contact with physics in high school. I also took the required mathematics courses in which I was a good student. If someone had told me that there were professional physicists and mathematicians who did this sort of thing for a living I would not have believed it. I had never met such a person. I had of course heard of Einstein but I had no idea of what he actually did. I also had no scientific curiosity. I tell you these things to explain what kind of student I was when I enrolled in Natural Sciences 3.

Cohen was a fine lecturer for this level of student. There were well over a hundred of us. He had a rich deep voice and a round reassuring handwriting when he wrote on the blackboards. We began with the Greeks, worked our way to Copernicus, Galileo and then Newton. Cohen was a Newtonian scholar, so we learned about Newton's life. For the first time I realized what a very strange man he had been. His masterwork, the *Principia* which he wrote at the end of the 17th century was presented in a way, deliberately, so that it would be accessible only to scholars. It was written in Latin and used geometric arguments which could have been replaced by much simpler demonstrations using calculus—which Newton had invented. Before it was published, Newton got into another nasty priority fight with Robert Hooke as to which one of them had first enunciated the correct law of gravitation. They had previously clashed in their work on light. On the other hand, Newton's *Opticks*, which was published in 1704, is a very different kind of book.

In the first place, it is written in English and the style is as congenial as Newton was capable of. It deals a great deal with the optical experiments Newton had performed, including the one with the prism that showed that white light was really a mixture of light of many colors. Unlike the *Principia* where Newton avoided speculation about the ultimate nature of things, in the *Opticks* he does speculate. There is the famous Query 31 where he talks about the atomic theory. He writes,

> "All these things being considered, it seems probable to me, that God, in the Beginning, form'd, Matter in solid, massy, hard, impenetrable Particles, of such Sizes and Figures, and with such other Properties and in such Proportion to Space, as most conduced to the End for which he form'd them, and that these primitive Particles being Solids, are incomparably harder than any porous Bodies compounded of them; even so hard, as never to wear or break in pieces; no ordinary Power being able to divide what God himself made one in the first Creation. While the Particles continue entire, they may compose Bodies of one and the same Nature and Texture in all Ages. But should they wear away, or break in pieces, the Nature of Things depending on them, would be changed. Water and Earth, composed of old worn Particles and Fragments of Particles, would not be of the same Nature and Texture now, with Water and Earth composed of entire Particles in the Beginning. And therefore, that Nature may be lasting, the Changes of corporeal Things are to be placed only in the various separations and new Associations and Motions of these permanent Particles, compound Bodies being apt to break, not in the midst of solid Particles, but where these Particles are laid together, and only touch in a few points . . . "

It is clear that when it came to material objects, Newton was an atomist *pure et dure*. But what about light? Here it seems to me that he was somewhat more nuanced. He writes, "In the first Books of these *Opticks*, I proceeded by the Analysis to discover and prove the original Differences of the Rays of Light in respect of Refrangibility, Reflexibility, and Colour, and their alternate Fits of easy Reflexion and easy Transmission, and the properties of Bodies, both opaque and pellucid on which their Reflexions and Colours depend...". I am not sure what meaning is to be attached to "alternate Fits of easy Reflexion and easy Transmission", but it does not sound like a simple particle picture to me, although this is how Newton was later interpreted. Opposed to this view was, in the first instance, that of Hooke who claimed that light was a wave. But of greater significance was the work of Christian Huygens—or Hugens—the great Dutch contemporary of Newton whom the latter came to visit in 1689, before Newton's *Opticks* had been published. I do not know if they discussed the nature of light. Huygens' own book was published in 1690. He proposed that light was an oscillation of waves in an aetherial medium. What he did not do was to consider the interference of waves. When two waves meet they can reinforce each other or interfere with each other destructively. Huygens' "waves" were actually pulses and not waves as we would think of them.[3] Nonetheless, either concept of light could account for the observations that were then available. Everything changed with the work of the late 18th and early 19th century British polymath Thomas Young. It is to him that we owe the idea that light waves can interfere with each other.

Over the years I have encountered a certain number of people that I would label "geniuses." I had an afternoon tea with Schrödinger in his apartment in Vienna and had a chance to show Dirac around that city. When I was in high school I spent some time with Duke Ellington. As I discussed, I had one memorable lunch with W.H. Auden and had the chance, when I was a Harvard undergraduate, to ask John von Neumann a question. The question was whether he thought the computing machine would ever replace the human mathematician. He response was, "Sonny, don't worry about it." What these people had in common was their ability to do things with apparent ease that most of us cannot do at all. By this standard, or any other, Thomas Young was a genius.

He was born into a Quaker family in Milverton in Somerset in 1773. By the age of fourteen, he knew not only Latin and Greek, but had a grounding in French, Italian, Hebrew, Chaldean, Syriac, Samaritan, Persian, Turkish and Amharic. Later in life he made basic contributions to the decipherment of Egyptian hieroglyphics, something that was completed by the French

---

[3] **I am grateful to the Newton scholar Alan Shapiro of the University of Minnesota for clarifying remarks on Newton, Huygens and Young.**

linguist Jean Françoise Champollion in 1822. Young, who became independently wealthy, was trained and practiced as a London physician and made some important contributions to the practice of medicine. This is not what made him immortal to physicists. Prior to the turn of the century, he began making experiments in physics. The ones that he performed on light were summarized in his November 1801 Bakerian lecture —"On the theory of light and colours"—before the Royal Society of which he had been a fellow since 1794. It was these experiments that persuaded most physicists that Huygens was right—light is a wave.

The experiment that he described in his Bakerian lecture is not the one that is usually explained to students. That came later. In this one, which was actually first done by Newton, he tells us that he made a square hole in a card. The breadth of the hole, he notes, was "66/1000" of an inch. To this hole he attached a hair that ran from top to bottom. It had a diameter of $1/600^{\text{th}}$ of an inch. Next he shone candle light on this arrangement. Nowadays we use lasers, which produce a concentrated beam. The hair split the light beam into two parts. I think of those bicycle races where the stream of riders is split in two by a traffic island in the middle of the road. We know that after they pass the island the two streams re-unite and to all appearances look as they did before. This is what the particle theory of light would predict. But this is not what Young observed. On the other side of the hair, light and dark fringes were observed. Using his candlelight, these were red fringes. Young understood immediately the implications of this. It had to do with the interference of light waves. Later he repeated this experiment with two pinholes instead of the hair and got similar results. This was the Young "double slit" experiment. While all of this was interesting to me as Cohen explained it, it was hardly life-changing. What happened next was.

It was getting close to the end of the first semester and Cohen presented a couple of lectures on the theory of relativity. I was overwhelmed. The idea that the speed of light was the universal speed limit and that every observer in a state of uniform motion would observe the same speed was incredible to me. There was also the claim that clocks in motion appear to run slower when observed by someone at rest. This also seemed incredible to me. What was the man talking about? In the course of this he said, I think as a joke although I did not see it that way, that only ten, or maybe it was twelve, people in the world understood the theory. Much later I learned that when the great British astronomer Arthur Eddington, who in the 1920s had written a noted monograph on relativity, he was asked if only three people in the world understood it. He replied, "Who is the third?" Eddington was certainly referring to Einstein's general theory of relativity and gravitation which at the time was not widely understood. In any event, with all the lunar audacity of a Harvard freshman I decided to become the eleventh or thirteenth person to understand the theory. I had no idea that there was a special and general theory. I had a plan, I would find a book that explained it.

Not only did I not have the remotest background, mathematical or otherwise, to understand the theory, I did not even know what it meant to understand something like relativity. In high school I had taken Spanish, so I knew what was meant by understanding a passage in Spanish. You translated it into English which presumably you "understood." If you came to a hard word you looked it up in a dictionary where it was "defined" in terms of words that you "understood." When it came to poetry what was meant by understanding was to be able to describe its meaning in terms of prose. In geometry what was meant by understanding was the capacity to prove something from the axioms, perhaps with the aid of diagrams. But what does it mean to understand the theory of relativity or the quantum theory for that matter? What it does not mean is to be able to "explain" the theory in terms of the theory that preceded it. If anything, this process should be the reverse. Be all that as it may, I went off to Widener Library to find a book.

I thought that I might as well get a book by Einstein since he surely understood the theory. There were a couple and I picked one called *The Meaning of Relativity*. I liked the title. It was the worst choice I could possibly have made. It was based on a series of lectures that Einstein delivered to specialists in 1921 at Princeton, although it has had revised editions. It is very sophisticated and highly mathematical. Apart from the opening sentence, "The theory of relativity is intimately connected with the theory of space and time.",[4] I understood essentially nothing. I might have given up but instead I went to Cohen. He might have dismissed me as just another foolish freshman, but he didn't. He told me that in the spring semester there would be a course on about the same level as his which would be devoted to modern physics. He told me that it would be taught by a man named Philipp Frank, who was actually a friend of Einstein's and had just written a biography of him. Cohen said that he would have no objections if I took that course along with his, so I signed up.

---

[4]**Princeton University Press, Princeton, New Jersey, 1950, 1.**

The course met once a week, I think it was Wednesday afternoons, for two hours. It met in the large lecture hall in the Jefferson physics laboratories. When I showed up for the first lecture the room was full. Later I learned that people from other universities in the area came to audit. I had no idea what to expect. What would Professor Frank look like? When I first saw him he seemed perfect for the part. He was a smallish man who walked with a bit of a limp. What hair was left was in kind of a halo around his head. His accent was not easy to place. Although he had been born in Vienna—in 1884—he had lived in Prague from 1912, when he succeeded Einstein at the German University, remaining until 1938. Then he came to the United States where Harvard cobbled together some kind of appointment. He knew all sorts of languages including Persian which he had studied in night school in Vienna. I envisioned his languages as being piled up on top of each other like the buried cities of Troy. When he spoke English there appeared the shards of the other languages. Professor Frank would lecture for an hour and then make what he called "a certain interval". After the "interval" he would answer questions or go into things with more depth, "If you know a little of mathematics."

His course also began with the Greeks. We learned that the Greek astronomers were committed to the notion that the celestial objects were attached to spheres that rotated uniformly. But this did not fit the facts. Planets, for example, periodically appear to move backwards. To account for this and still to "save the appearances" the uniformly moving planets moved in spheres around spheres—epicycles. Cohen had discussed this but Professor Frank added a new insight. Kepler had replaced all of this with single elliptical orbits. Professor Frank noted that such a motion can always be reproduced as a series of uniform circular motions so the two models cannot be distinguished if all one is interested in is describing the observed motions. Other criteria enter such as simplicity or "elegance." When it came to Newton, Professor Frank explained the story of the falling apple. He said to imagine an apple on a tree that grows to be as long as here to the Moon. Then the Moon is an apple on the tree and it too must fall in the field of the Earth's gravity. Then we came to relativity.

Professor Frank explained the relativity principle which was first stated by Galileo; the laws of physics are the same for observers at rest and observers moving uniformly with respect to these observers. Indeed, this second group of observers can perfectly well claim that they are the ones who are at rest. He told us that Einstein from very early on believed that this must apply to all laws of physics but he realized that there was a contradiction. If he was allowed to move with the speed of light he could ride on a light wave. Then it would no longer look like a wave, so he would know that he was moving at the speed of light, which violates the relativity principle. It took him many years from his having first thought of this paradox to resolving it. You cannot in relativity move with the speed of light. Professor Frank had a simple way of explaining the transformations that take you from one system of coordinates to another that is moving uniformly. I have used it often when teaching the subject. He also told a funny story. He visited Einstein in Berlin a few years after Einstein had published his popular book on relativity. By this time Einstein had married his second wife and was living with his two step-daughters, one of whom was present. Einstein was explaining how his book was so simple that even his step-daughter could understand it. He then left the room and Professor Frank asked her if it was really true that she understood it. Yes, she said, she understood everything except what a coordinate system was.

Following relativity, Professor Frank turned to light. He told us about the double slit experiment of Young and he noted that if Young had been able to observe them he would have seen wave-like manifestations even from a single slit. There would be a build-up of intensities in front of the slit, but then there would be fewer peaks at spaces to either side. The profile would look like that of a mountain with its satellites to either side. All of this showed that in these experiments, light behaved like a wave. But then Professor Frank told us about one of Einstein's 1905 papers, actually the one for which he won the Nobel Prize, in which he argued that in other experimental arrangements, light would behave like a particle. In particular if you shone light on a metallic surface, electrons would be liberated from the metal—the photoelectric effect. This would also be predicted from classical physics. But what Einstein was proposing was that the energy of these liberated electrons did not depend on the intensity of the incident light, but only on its color. The more violet the light, the more energetic were the liberated electrons. In these experiments light behaved like particles—quanta—whose energy was proportional to the frequency of the radiation. Professor Frank said that if the intensity of the light was diminished so that it consisted of a single quantum, this quantum could still supply the energy to liberate an electron. What is light? It is a wave or a particle? It is both and neither. Then, we turned to electrons.

Professor Frank told us about the proposal of Louis de Broglie, that electrons exhibited, under certain conditions, a wave character. He told us about the experiments—analogous to those of Young—in which electrons were made to impinge on a grating of slits. They were then deposited on a detector and one could observe the same kind of lines of intensity that Young had observed for light. What were electrons—waves, particles, both, neither? Electrons, Professor Frank said, are electrons. But then he suggested a thought experiment—one that he said had never been carried out—it has now.[5] Suppose you had a double slit and that your beam of electrons was so weak that only one electron at a time went through the system. What would happen? If both slits were open, Professor Frank told us, then over the course of time—at least according to the quantum theory—the single electrons would re-recreate the whole pattern. The pattern would not be, as one would expect for particles, the sum of the two open slit patterns. It would show the whole effect of the interference. There was no way of predicting where a given electron would land but you could predict, using the theory, where it was most likely to land. This to me was already amazing, but what he said next was more than amazing. It ultimately convinced me to go into physics. Suppose you closed one of the slits and allowed the electrons to enter the apparatus one after the other; what would happen? The two-slit pattern would disappear and be replaced over the course of time by the one-slit pattern. The question that immediately occurred to me was, how does the single electron going though the slit "know" that the other slit has been closed? Where does that information come from? By now I know that quantum mechanics does not provide an answer. In fact, it is not even the right question to ask, but at the time I didn't know this and was completely baffled and became determined to learn more and so I began a long course of study so that I eventually ended up with a Ph.D. in physics.[6]

I have often wondered if young people who are now first exposed to the quantum theory have the same sense of wonderment that I did. After all, when I first learned it, the theory was only about twenty years old. All of the creators were still around lecturing and writing. Now the theory is well over a half-century old. Do young people simply accept the rules and get on with it? What I do know is that some non-physicists do have this sense of wonder—even if, or mainly if, they do not really understand the

[5] See for example http://en.wikipedia.org/wiki/Double-slit_experiment for a review with many additional references.

[6] A reader who wants to know the details of this improbable saga can consult my autobiographical memoir *The Life it Brings*, Penguin, New York, 1988.

theory. But they have absorbed enough so that it has entered their creative imagination. A very interesting example is the playwright Tom Stoppard.

In the interests of full disclosure once again, I have to confess that I am a Stoppardian of the deepest dye. Plays like Jumpers or Travesties are for me intellectual champagne. The mixture of ideas and word-play are for me irresistible. In the fall of 1994 I received a phone call from a woman named Anne Cattaneo. She identified herself as the editor of a journal named *The New Theater Review*—now called *The Lincoln Center Theater Review*. She told me that the Lincoln Center was mounting a play by Stoppard called "Hapgood" which had premiered in London in 1988. She also told me that the play involved quantum mechanics in some way. I had read about such a play and been intrigued. Stoppard and quantum mechanics—what a pairing. Then she went on to tell me that *The New Theater Review* was planning an issue devoted to the play and its scientific implications. Would I like to contribute? She said that Stoppard was going to be one of the contributors. The prospect of sharing the stage with Stoppard was irresistible, so I agreed. The first thing that I did was to buy a copy of the play.

"Hapgood" is both the name of the play and the name of the principal *character*. Her first name, which is almost never used in the play, is Elizabeth—Lilya in Russian. She is a spy, indeed, the head of a coven of spies who refer to her as "mother." They are British except for one, Joseph Kerner, who is Russian. It was Hapgood who turned him but, in the course of things, the turning got a bit out of hand. Hapgood became pregnant and gave birth to a son, Joe. He is an adolescent and does not know who his father is. He is much more interested in playing rugby. Some of the play takes place during his school games. Kerner works at CERN, the elementary particle laboratory in Geneva. He is working on an anti-ballistic missile warhead that involves anti-matter. This is only marginally more absurd than other anti-ballistic missile programs I have heard about. There is a convoluted plot involving Kerner giving some or all of this information to the Russians. Periodically, Kerner gives small lectures about quantum theory and its relevance to human behavior. Schrödinger had enough trouble applying quantum theory to cats let alone human beings. I will shortly deconstruct one of these lectures, but first let me say something of how Stoppard appears to have gotten into this. What I know is what he wrote in the *The New Theater Review*—an article which he entitled "The Matter of Metaphor." My article, which followed his, was called

"A Trick of Light" and covered some of the same ground, although in considerably less detail, than what I have been discussing here. According to Stoppard, "...Science for non-scientists is a boom area in publishing, and Hapgood is itself a fruit of a dozen books about quantum physics written for a general readership. If there were ever a general reader, I am one. I am not even a closet scientist, and certainly not a frustrated scientist. In fact, I probably owe it to my general lack of scientific education that the central image of Hapgood—the dual nature of light—excited me so much when I finally caught up with what is wearily familiar to anyone who stuck with physics past high school."[7]

"Wearily familiar" does not describe my experience as I have tried to explain.

Stoppard goes on, "But, of course, it excited me for its potential as a metaphor. Hapgood is not "about" physics, it's about dualities. No—let the playwright correct the critic in me—Hapgood is not about dualities, of course, it's about a woman called Hapgood and what happened to her between Wednesday morning and Saturday afternoon 1989 just before the Berlin Wall was breached."

Stoppard continues, "As a matter of fact, the story has much more to do with espionage than physics, but I won't deflect any compliments that might be going for a play with a reasonably plausible physicist on board, because the springs of the play are indeed science; it was only in looking around for a real world metaphor, that I hit upon the le Carré world of agents and double agents. The physics came first, the woman called Hapgood came second."

And he adds, "Incidentally, while several of the physicists who saw the play went out of their way to be kind about its general veracity, John le Carré, whom I came to know and, I seem to recall, who went on my tickets, managed to avoid mentioning the play from that moment on; which I immediately recognized as the height of courtesy... ".[7]

Stoppard does not tell us which physicists vouched for the play's "general veracity" or whether they both saw it and read it. Things go by pretty quickly in an actual performance. Nor does he tell us which popular books he read. It would be interesting to know. We do know, since he quotes from it, that he looked at Feynman's lectures.[8] This is a bit like my consulting Einstein's Princeton lectures to learn about relativity. Feynman gave

[7] ***The New Theater Review*, 4.**

[8] **R.P. Feynman, R.B. Leighton and M. Sands, *1963 Feynman Lectures.***

these lectures to undergraduates at the California Institute of Technology—Caltech. Being Feynman, they are full of original insights. They are highly technical. I was told that the undergraduates found them pretty tough-going although the faculty members, who audited them in droves, loved them. It is the last place I would send a non-scientist who wanted to get a flavor of quantum theory. I was also told that Stoppard visited Caltech and actually sat for awhile in Feynman's empty office—osmosis. What I want to do is to deconstruct one of Kerner's mini-lectures on quantum theory. I do this in the spirit of the seminars that Bohr used to moderate, where his many interruptions of the speaker were always accompanied by his saying that he was not there to criticize, but only to learn.

Kerner is talking to Paul Blair, another agent, who has informed Kerner that his cover as a double agent has been "blown." Kerner's career is over "except as a scientist." Kerner makes the point that a double agent is not like a giraffe—something definite but "more like a trick of the light." Then he explains,

> "KERNER: Look (He points.) Look at the edge of the shadow. It is straight like the edge of the wall that makes it. This means that light is particles: little bullets. Bullets go straight. They cannot bend round the wall a little, like water bends round a stone in the river."

I am surprised that none of the physicists Stoppard had contact with had informed him that this characterization of light is entirely wrong. It had been observed as early as the 17th century that shadows cast by light were fuzzy. Hooke and Huygens argued that this implied that light had a wave nature which manifested itself in this kind of observation. Light waves can interfere with each other whereas particles, as we usually conceive of them, cannot. When Einstein wrote his 1905 paper claiming that under some circumstances light could exhibit particle behavior, he began it by explaining how such a thing was possible given all the evidence that it was a wave. Thus began the idea of the complementary nature of light.

Blair responds;

> "BLAIR: (Irritated) Yes. Absolutely.
> KERNER: So that's what. When you shine light through a gap in the wall it's particles. Unfortunately, when you shine light through two little gaps, side by side, you don't get particle pattern like for bullets, you get wave pattern like for water. The two beams of light mix together and..."

In what I have written above, I have explained that if you do the one-slit experiment for light, or electrons for that matter, you get a pattern of light and dark places on the detecting screen. This happens because different parts of the slit produce waves that interfere. Even in the one-slit case, the sort of question that bothered me as a freshman arises. Suppose you let the electrons through one at a time. Each electron will land at some spot on the detector, but as they accumulate they reproduce collectively the pattern of light and dark lines. How does the single electron know how to do this? How does it know where to go to play its role in the final pattern?

Conventional quantum mechanics has an answer, but it is not one that would have satisfied me as a freshman. By now I am used to it. The equation of Schrödinger has a solution which is a function of space and time. It you take this solution and square it, this gives the probability of finding the electron, say, at some place at a given time.[9] Where the function is large the probability of finding the electron is greatest. Nothing in the theory tells you where a given electron will land, only where it is likely to land. This is one of the things that troubled Einstein. It also troubled David Bohm, who produced his alternate interpretation of the quantum theory with its guide waves. The answer to the question of how an electron knows what to do in Bohmian mechanics is that the guide wave tells it. Back to the play.

> "BLAIR: Joseph, I want to know if you're ours or theirs, that's all.
> KERNER: I'm telling you but you're not listening. Now we come to the exciting part. We will watch the bullets to see how they make waves. This is not difficult, the apparatus is simple. So we look carefully, and we see the bullets, one at a time. Some go through one gap and some go through the other gap. No problem. Now we come to my favourite bit. The wave pattern has disappeared. It has become particle pattern again."

In 1927 there was a conference of leading physicists in Como, Italy to celebrate the centenary of the death of Alessando Volta. Quantum mechanics was something like two years old. Heisenberg had just published his uncertainty principle. Both Bohr and Einstein were at the conference. Bohr introduced his notion of complementarity. Einstein had a little surprise for him. It was a thought experiment involving double slits. But these were

[9] **Technically, the wave function is a so-called complex number so you must take the square of what is called the absolute value.**

mounted on rollers or wheels which made the apparatus movable. Now suppose you fix on a place on the detector above both slits. An electron comes to the lower slit, say. To reach your position, the electron which has come through the slit must change its direction, which means that it must change its momentum. Since momentum is conserved the apparatus acquires this momentum in reverse—it recoils—and that can be detected since it will move on its wheels. This recoil momentum can be measured. If the electron comes through the upper slit, there will also be a momentum change but less, since the angle it goes through is smaller. Hence by measuring the momentum recoil of the apparatus, we can tell through which slit it went, so it would appear that in one and the same experimental arrangement we can determine both the particle and wave aspects of the electron. The wave aspect would be manifest by the intensity patterns of electrons on the detector. But Bohr had an answer. The recoil momentum of the apparatus must be measured to a certain accuracy in order to make this determination. But by Heisenberg's uncertainty principle, this limits the accuracy of the position of the slits. If you work out the details this inaccuracy is greater than the distance between the first and most prominent maxima in the interference pattern. The slits are effectively as wide as the interference pattern so there is no interference pattern and Einstein's thought experiment does not work. Sometimes one reads in popular books that this set-up shows that the electron can be in two places at once. The electron is in a "place" only when you observe it on the detecting screen. If someone tells you that it is also in another place, the proper answer is "show me." Recall Kerner's remark I quoted at the beginning: "An electron can be here or there at the same moment." In so far as it is "here or "there", it is either one or the other—not both. The play goes on.

> "BLAIR: How?
> KERNER: Nobody knows. Somehow light is continuous and also discontinuous. The experimenter makes the choice. You get what you interrogate for ... ".[10]

Blair's question, "How", is exactly the kind of question I asked as a freshman and to which there is no answer, at least in the quantum theory. Stoppard's quantum theory reminds me of a conversation I had with Marvin Minsky when we were both undergraduates. He later went on to become one of the founders of artificial intelligence. He showed me a drawing he had

[10]**Hapgood, 500 – 1.**

made of a bird. "This looks like a bird," he said, "but no bird looks like this." Stoppard's physics looks like quantum mechanics, but no quantum mechanics looks like Stoppard's physics.

Professor Frank died on the 21st of July 1966. Later that year there was a memorial for him at Harvard at which I was one of the speakers. His widow Hania, whom he had married in Prague when she was one of his students, was there. She belonged to an intellectual circle which included Kafka. She was a remarkable and very lively woman. I have an ineluctable memory of calling one night to speak to Professor Frank. Hania answered and in her inimitable accent, said, "We are here singing English folk songs and Philipp has gone away."

Chapter 5

A Measurement

"We wish to measure a temperature. If we want, we can pursue this process numerically until we have the temperature of the environment of the mercury container of the thermometer, and then say: this temperature is measured by the thermometer. But we can carry the calculation further, and from the properties of the mercury, which can be explained in kinetic and molecular terms, we can calculate its heating, expansion, and from the resultant length of the mercury column, and then say: this length is seen by the observer. Going still further, and taking the light source into consideration, we could find out the reflection of the light quanta on the opaque mercury column, and the path of the remaining light quanta into the eye of the observer: their refraction in the eye lens, and the formation of an image on the retina, and then we would say: this image is registered by the retina of the observer. And were our physiological knowledge more precise than it is today, we could go still further, tracing the chemical reactions which produce the impressions of this image on the retina, in the optic nerve tract and in the brain, and then in the end say: these chemical changes of his brain cells are perceived by the observer. But in any case, no matter how we calculate—to the mercury vessel, to the scale of the thermometer, to the retina, or into the brain, at some time we must say: and this is perceived by the observer. That is, we must always divide the world into two parts, the one being the observed system, and the other the observer. In the former, we can follow up all physical processes (in principle at least) arbitrarily precisely. In the latter this is meaningless. The boundary between the two is arbitrary to a very large extent. In particular we saw in the four different possibilities in the example above, that the observer in this sense needs not to become identified with the body of the actual observer: In one instance in the above example, we included even the thermometer in it, while in another instance, even the eyes and optic nerve were not included. That this boundary can be pushed arbitrarily deeply into the interior of the body of the actual observer is the content of the principle of the psycho-physical parallelism—but this does not change the fact that in each method of description the boundary must be put somewhere, if the method is not to proceed vacuously, i.e.; if a comparison with experiment is to be possible. Indeed experience only makes statements of this type: an observer had made a certain (subjective) observation; and never any like this: a physical quantity has a certain value."

John von Neumann[1]

"The idea that consciousness collapses the quantum wave was originally proposed by John von Neumann in the 1930's. What took us so long to take this idea seriously?"

Amit Goswami[2]

When I finished taking Philipp Frank's course on the foundations of modern physics in the spring of 1958, one thing had become clear to me: if I really wanted to have any kind of understanding I needed to learn some mathematics. Professor Frank, after his formal lecture, made as I said in the last chapter, "a certain interval", which was followed by a question period during which he would have a chance to expand on some things "for those of you who know a little of mathematics." I still remember his solving the elementary differential equation that described the decay of a particle. I had no idea what he was doing or how it led to the exponential answer. I also had no idea what I was doing except that I wanted to learn "a little of mathematics." On how I would use this later in life, I was clueless. I had some friends who were taking the introductory course in calculus. Indeed, one of them showed me his homework papers. The grader had made some extremely witty remarks. It turned out to have been Tom Lehrer who became one of the foremost musical satirists of the 20th century.

These friends spoke of "infinitesimals", an essential feature of the calculus, as if they represented the limit of possible human understanding. Grasping them, they said, bordered on the impossible. This worried me, so I went to my freshman advisor and asked if he could look up my mathematical aptitude scores from the tests I had taken to get into Harvard in the first place. He did and said that he thought that it would be alright for me to take the calculus. In the fall of my sophomore year, I duly enrolled. In the event I quite enjoyed the course, infinitesimals not withstanding. What I did not enjoy was freshman physics, which I took as a sophomore. I had by no means committed myself to physics as a major, but was testing

[1]**John von Neumann, *Mathematical Foundations of Quantum Mechanics*, Princeton University Press, Princeton, New Jersey, 1955, 419 – 20.**

[2]**Amit Goswami, *The Self-Aware Universe*, Penguin Putnam Inc, New York 1993, 205.**

the waters. I found the course which dealt with the basics, such as pulleys and levers, inclined planes, and the rest, exceedingly dull. I realized that Professor Frank's course was the dessert and that I was now dealing with the entrée. To make matters worse, there were laboratory sessions which I hated. I have told people over the years that the reason for my problems in the lab was myopia. I could not read a slide rule. But the real reason was that I had no mechanical aptitude. Nothing ever worked right. The whole experience has given me a great deal of respect for experimental physicists—indeed, experimental anything. But after this course I never took another undergraduate course in physics.

I now took math courses both during the academic year and in the summers. That became my major and I acquired a tutor—the late George Mackey. Mackey was a wonderful teacher who, at the time, lived a sort of monk's existence devoted to mathematics. Among other things he was bringing up to date the mathematical foundations of the quantum theory—a more modern version of von Neumann. I learned from Mackey what sort of mathematics was used in the quantum theory and sort of headed in that direction. Up to this point I had never taken a course on quantum theory. The mathematics department at Harvard had a kind of policy of discouraging Harvard undergraduates to apply to the graduate school. They thought it was a good idea for students to have a new experience. You had to have at least a *magna cum laude* for them to consider your application. By this time, I had come to like life in Cambridge a great deal and as I had graduated *magna* I applied and was accepted. One of the first things I did was to take the graduate course in the quantum theory.

The course was taught by Julian Schwinger, the star theoretical physicist at Harvard. He was one of the creators of quantum electrodynamics for which, along with Feynman and Shinichiro Tomonaga, he shared the 1965 Nobel Prize in Physics. There was a sort of theater in Schwinger's lectures. He lectured seamlessly without notes and you were not supposed to ask questions. The first semester was largely devoted to why the quantum theory was necessary and to things like the uncertainty principle. There was no discussion of the Einstein–Podolsky–Rosen paper, and when Bohm's papers came out at about this time, I once heard Schwinger say that he thought that they were pointless. After this introduction, Schwinger launched into his own formalism for the theory which left most of the class, including me, rather lost. We used to go down to MIT where Victor Weisskopf was giving a much more homespun version to fill in the gaps. After two years I received a masters degree and it became time to seriously think about the future.

By this time I had gotten to know a number of mathematicians, including my fellow graduate students. It had become clearer and clearer to me that I did not have the abilities to do creative mathematics at a level that I would find satisfying. Unless you have been involved in one of these creative disciplines this is difficult to explain. In my case you could point to my course record which was very good and say wasn't that proof enough. But I could see that my way of understanding things was different from the way the truly gifted students understood them. I could follow proofs but for them the proof was almost a secondary matter since the truth of the theorem was to them self-evident. While I was having this self-doubt, something happened that settled the matter. Garrett Birkhoff, who was the chairman of the mathematics department, called me in for a chat. He had noticed that I had been taking courses in classical mechanics, methods of mathematical physics and Schwinger's quantum mechanics. He said that it seemed to him that I was more interested in physics than mathematics and that I would have to choose. I decided that I would try to switch to physics.

Armed with my transcript, I went to the chairman of the physics department and asked to make the switch. He looked over my record and drew the opposite conclusion to that of Birkhoff—namely I had taken too few physics courses. Apart from freshman physics and my graduate courses I had taken none. On the other hand I had excellent grades, so I was accepted with two conditions. The first was that I was to spend the summer working at the Harvard Cyclotron Laboratory to get some real hands-on experience with experimental physics, and the second was that I was to take an oral exam on general physics during the next year. I agreed to both.

The first Harvard cyclotron was completed in 1938. When the atomic bomb project moved to Los Alamos, the cyclotron was sold to the government for one dollar with an understanding that the government would try to be helpful in constructing a new one after the war. Indeed, this is what happened, and with the help of the navy the new cyclotron went into operation in 1949. By present day standards it was almost a toy. It was simple enough so that faculty and graduate students could operate it. No ordinary physicist would be allowed to get anywhere near the control room of one of the large modern accelerators. It would be like inviting a flight attendant on a trans-Atlantic airliner to come into the cockpit and fly the plane. In my case, the faculty member in question was a new assistant professor named Walter Selove. He was doing an experiment in nuclear physics and my job was to assist him. But what I actually did had nothing to do with physics. I piled stacks of lead bricks for shielding and fitted various targets to be bombarded by the protons in the beam. I had no idea what the experiment was trying to prove. In short, while I had a good time that summer, I learned nothing.

The following fall I studied for my forthcoming oral exam and, as strange as it may sound, I had begun work on what became my PhD thesis. I had been able to do this because there was another young faculty member named Abraham Klein who had a project which had variants of the kind that made theses. They did not require a deep understanding of physics but rather the patience to carry out some fairly involved calculations. I think I was Klein's first graduate student. He and Walter Selove went on to have distinguished careers at the University of Pennsylvania. But there was the oral exam. My performance was distinctly odd. I recall one question in which I was asked to show why a certain hypothetical particle decay could not happen. I gave an arcane argument involving something called "charge conjugation" whereas the argument they were looking for involved the conservation of energy and momentum—something that any real physicist would have thought of at once. I did not exactly fail the exam but I did not exactly pass it either. They would have been within their rights to ask me to repeat it, but instead they asked me to take a course in modern experimental physics.

The course was given by Robert Pound, a noted experimental physicist and a very nice man. Maybe he had been tipped off as to what I was doing there. He treated me very well, including assigning to me a lab partner named Paul Condon. Condon was the son of the very well-known physicist E.U. Condon, so I suppose physics was in his genes. His brother Joe was also a physicist. Paul and I soon worked out an amiable *modus operandi.* I was allowed to comment but not to touch any of the apparatus. I was also allowed to peek at his lab book, which was a model. Pound would come around to our lab bench from time to time and offer encouragement. One of the experiments we did was to measure the spectrum of radiation in a heated cavity—so-called "black body radiation." It was the report of new measurements of this spectrum at the dawning of the 20th century that led Max Planck to introduce the notion of the "quantum"—his name. The only way that Planck could explain this distribution was to imagine that the electrons attached to the atoms in the walls of the cavity were set into vibration when the walls were heated. These oscillating electrons emitted and absorbed radiation.

To produce his distribution Plank had to assume that these oscillators had a restricted set of energies. In our language we would say that the energies were "quantized." Professor Frank once noted that it was like beer being sold in pints and quarts—the quanta. Planck introduced a constant that set the scale of these energies. "Planck's constant" is one of the basic natural units. In our modest experiment we could actually measure this constant. Einstein transformed Planck's idea in 1905 when he argued that the radiation in the cavity was made up of quanta. Professor Frank said it was like the requirement that whenever you found any volume of beer it was always broken up into pints and quarts. In any event, with the help of Condon, I passed the course and was allowed to complete my thesis. It never occurred to me at the time what the philosophical implications of our black body experiment—and some of the others we did—was. We were actually a living realization of the basic doctrine of Bohr about the nature of quantum mechanical measurements.

In our experiment we never dealt with the actual quantum processes that were taking place within the cavity. What we did was to read various dials and pointers and the like. These dials and pointers were completely classical, which is why we could discuss them in everyday language. Bohr insisted that quantum measurements were always like this. At the end of the day there were observations that could be described in ordinary language since this was the only language we knew. Some division had to be made between the underlying quantum process and the classical world of the observer. He wrote about this again and again. Here is a typical example:

"This necessity of discriminating in each experimental arrangement between those parts of the physical system considered which are to be treated as measuring instruments and those which constitute the objects under investigation may indeed be said to form a principal distinction between classical and quantum-mechanical description of physical phenomena. It is true that the place within each measuring procedure where this discrimination is made is, in both cases, largely a matter of convenience. While, however in classical physics the distinction between object and measuring agencies does not entail any difference in the character of the description of the phenomena concerned, its fundamental importance in quantum theory, as we have seen, has its root in the indispensible use of classical concepts in the interpretation of all proper measurements even though the classical theories do not suffice in accounting for the new types of regularities with which we are concerned in atomic physics ... ".[3]

In his discussions of the foundations of the quantum theory, John Bell introduced the invaluable concept of FAPP—For All Practical Purposes. For most physicists, most of the time when it comes to quantum mechanics, FAPP is perfectly good. We may have to dig deeper if we have to teach a course on it and use one of the modern textbooks that discuss such things as the Einstein–Podolsky–Rosen experiment and Bell's inequality. From a FAPP point of view a statement like the one of Bohr's above seems quite reasonable. But it is the sort of thing that drove Bell to distraction and caused him to label Bohr as an "obscurantist." Nowhere in Bohr's writings could he find any clear criteria for making the division Bohr talks about between the quantum and classical realms. Bohr appears to maintain that it is somehow a division between large and small things—dials and pointers versus atoms. But this clearly does not work. Take, as an example, the thought experiment that Einstein came up with in 1927 to refute Bohr's notion of complementarity, the one I described with the slits on wheels. It is hard to think of anything more classical. But to resolve the paradox that Einstein presented, one must invoke the uncertainty principle. It is hard to think of anything more quantum mechanical. Where in this experiment has the classical realm stopped? How do we know in principle—and not FAPP—where to make this division? You will not find the answer in Bohr. You will also not find any discussion of what exactly is a quantum mechanical measurement. In particular, can you describe one using the formalism you might find in an average textbook?

The first person who seems to have discussed this in any generality was von Neumann in his 1932 book on the mathematical foundations of the theory. He made a distinction between what he called two types of "intervention" which can occur in a quantum mechanical system. One type of intervention is describable by the ordinary Schrödinger equation. We begin by specifying the Schrödinger wave function at some time and place. The system is acted on by a force—a magnetic field, say. The wave function continually evolves according to the Schrödinger equation until at some later time and at some other place it acquires a new value. If we want, we can use these values as the initial conditions and run the scenario backwards in time. Using the term of art, the equation is "time reversible." But, so far, we have not performed a measurement. To see what is involved, let us consider a specific example—the measurement of spins using magnets.

---

[3] **Niels Bohr, *The Philosophical Writings of Niels Bohr*, Volume IV, Ox Bow Press, Woodbridge, Connecticut, 1998, 81.**

Suppose we have some source that produces spin ½ particles such as electrons. A beam of them is produced by the source that has a mixture of spin up and spin down with proportions given by numbers a and b. The spin part of the wave function we can write as a(↑) + b(↓). Until these particles encounter the magnet, the wave function propagates as described by the Schrödinger equation. But we know from experiment that once the beam interacts with the magnet it is divided into two components, one with the spin up particles and one with the spin down. This process is still reversible but once a particle in one of the beams is actually detected, the situation is drastically altered. The question is, can this actual measurement be described by an ordinary time reversible Schrödinger equation? The answer is "no." Once a particle has been measured in the detector with, say, spin up, its wave function is spin up. The spin down part is gone. This is often referred to as the "collapse of the wave function." It seems more like a decapitation. It cannot be described by the usual Schrödinger equation. There are various ways of proving this depending on the level of mathematical sophistication one wants to bring to bear. The point is that this measurement is not time reversible. You cannot use the Schrödinger equation to reconstruct the *status quo ante.* All the king's horses and all the king's men will not create the old wave function again. In fact, you cannot reconstruct the past which puts history in an odd light. All you can do is to offer different pasts with different probabilities. Von Neumann does not discuss this. He simply states as a postulate that you need a different kind of intervention to carry out measurements, and he presents their mathematical properties. This is the "measurement problem" and we have been living with it ever since.

Veritable forests have been cut down to produce the paper that has been expended on this. I will simply give a personal overview. I state from the beginning that I do not know what the solution is. I will divide my discussion into two parts. The first part will deal with the class of proposed solutions in which the wave function does not collapse and the second part deals with the class of proposed solutions in which the wave function does collapse and an explanation is offered. Each of these classes has its proponents and critics.

The first example in the first class is Bohmian mechanics. In the first of his two papers,[4] Bohm discusses an example of what is now fashionably called "decoherence." It will be recalled that the Schrödinger equation

[4] ***Physical Review*, 85, 166–93, (1952).**

in Bohmian mechanics is used to derive the guide waves for the classical particles. What Bohm states is that when the entangled system enters a region where there is a perturbation due to the interaction with a measuring apparatus, there is a very rapid exchange of energy between the parts of the entangled function which causes it to break up into separable wave packets—decoherence. These separated packets guide the particles to the various parts of the detector. The wave function does not collapse but rather breaks up into pieces. It is shown that this produces the quantum mechanical answer.

Probably, the most popular version of the non-collapsing situation goes under the names "many worlds", "many paths" or "many histories." It dates to a PhD thesis written in 1957 at Princeton under the supervision of the late John Wheeler, by Hugh Everett III. The thesis was called "The Theory of the Universal Wave Function". In 1959, Everett published a paper on it with the more austere title, "Relative State Formulation of Quantum Mechanics",[5] a name suggested by Wheeler. As far as I know, this is the only thing Everett ever published on this subject. Wheeler was, at least at first, very enthusiastic about it. He arranged for Everett to visit Bohr in Copenhagen. Neither Bohr, nor anyone else at the time, showed any interest in Everett's interpretation. He then got out of physics and made a great deal of money as a defense analyst. He died of a heart attack at the age of 51 in 1982. By this time there was considerable interest spearheaded by the physicist Bryce Seligman DeWitt. In the 1970s he wrote both technical and popular articles about it and coined the name "many universe" interpretation. It is now more often called the "many worlds interpretation." What is it?

It has in common with the Bohmian mechanics interpretation that entangled wave functions decohere when the system encounters some sort of perturbation. This could be, for example, the photons left over from the Big Bang. Measurements are just another kind of perturbation. In Bohmian mechanics the different branches of the wave function act as different guide waves for the classical particles. In the Everett interpretation a wave function is a wave function and it is a solution at all places and at all times of the Schrödinger equation. But we only follow one of the branches. The others remain, and have nothing to do with us. They belong to another world with which we have no contact. If we use a Stern–Gerlach magnet to measure a spin and find a spin up, the spin down part of the wave function

[5] ***Rev.Mod.Phys*, 29, 454 – 62.**

is not lopped off but represents another possible outcome which is not our concern. What has been the concern of the physicists who have rejected this interpretation is, among other things, the proliferation of "worlds". A world tree must acquire so many branches that it begins to resemble a forest. This is an aesthetic objection, but there is also a physics objection—Born's rule. Returning to our Stern–Gerlach example, recall that the coefficients a and b in the wave function determine the probability that an experiment will yield spin up or down. Experiment confirms this assignment. Any interpretation of the quantum theory must yield this rule. It is all very well to say that we are on one branch or the other of the Everett many worlds, but what is the probability that we find ourselves on this branch and not that one—Born's rule. There has been much work on this matter by the practitioners but, as far as I can tell, it is still an open question. I now turn to the other possibility, that the wave function does collapse.

Whatever method one invents to get the wave function to collapse must go beyond the ordinary Schrödinger equation. Some new physics must enter. One possibility is that there is a wave collapsing force that acts randomly but just in the right places and times to do the job without producing conflicts. This has been explored in some detail. It would be more convincing if some of the observable consequences are ever observed. Another avenue has been the claim that it is human consciousness that acts to collapse the wave function. Before one dismisses this as so much New Age flapdoodle, one should at least consider some of the sources. Von Neumann seemed to hint that this is what he believed. His long-time friend and fellow Hungarian, Eugene Wigner, who was one of the most important theoretical physicists of the 20th century, was absolutely convinced. When he was awarded the Nobel Prize in 1963, he apparently asked his caller from Stockholm for which of his achievements he had been given the prize. I doubt that it was for his 1961 article, which he called "Remarks on the Mind-Body Question."[6]

Wigner makes his position clear right from the outset. He writes,

> "... When the province of physical theory was extended to encompass microscopic phenomena, through the creation of quantum mechanics, the concept of consciousness

[6]**I,J. Good ed., *The Scientist Speculates*, I,J. Good, 284 – 302, Heinemann, London, 1961. It is reprinted in *Quantum Theory and Measurement,* edited by J.A. Wheeler and W.H. Zurek, Princeton University Press, Princeton, New Jersey, 1983. The page numbers I will give refer to this book.**

> came to the forefront: it was not possible to formulate the laws of quantum mechanics in a fully consistent way without reference to consciousness. All that quantum mechanics purports to provide are probability connections between subsequent impressions (also called "apperceptions") of the consciousness, and even though the dividing line between the observer whose consciousness is being affected, and the observed physical object can be shifted towards the one or the other, it cannot be eliminated. It may be premature to believe that the present philosophy of quantum mechanics will remain a permanent feature of future physical theories, it will remain remarkable in whatever way our future concepts may develop, that the very study of the external world led to the conclusion that the content of the consciousness is an ultimate reality."[7]

One may well wonder, if humans had not fortuitously taken over from pterodactyls, whether the collective consciousnesses of pterodactyls would have done the trick. Be that as it may, Wigner goes on to a discussion which has become as celebrated in these circles as Schrödinger's cat. I will quote what Wigner writes and then comment. He begins with a general statement,

> "...In general there are many types of interactions into which one can enter with the system, leading to different types of observations and measurements. Also, the probabilities of the various possible impressions gained at the next interaction may depend not only on the last, but on the results of many prior observations. The important point is that the impression which one gains at an interaction may, and in general does, modify the probabilities with which one gains the various possible impressions at later interactions. In other words, the impression which one gains at an interaction, called also the result of an observation, modifies the wave function of the system. The modified wave function is, furthermore, in general unpredictable before the impression gained at the interaction has entered our consciousness: it is the entering of an impression into our consciousness which alters the wave

[7]**Wigner in Wheeler *et al,* 169.**

> function because it modifies our appraisal of the probabilities for different impressions which we expect to receive in the future. It is at this point that the consciousness enters the theory unavoidably. If one speaks in terms of the wave function, its changes are coupled with the entering of impressions into our consciousness. If one formulates the laws of quantum mechanics in terms of probabilities of impressions, these are *ipso facto* the primary concepts with which one deals."[8]

But Wigner goes further. It is this part which has been seized on by the multitudes—Wigner's friend. He writes, "It is natural to inquire about the situation if one does not make the observation oneself, but lets someone else carry it out. What is the wave function if my friend looked at the place where the flash might show at time t? The answer is that the information about the object cannot be described by a wave function. One could attribute a wave function to the joint system: friend plus object, and this joint system would have a wave function also after the interaction, that is after my friend has looked. I can then enter into interaction with this joint system by asking my friend whether he saw a flash. If his answer gives me the impression that he did, the joint wave function of friend + object will change into one in which they even have separate wave functions ... and the wave function of the object is ψ_1. If he says no, the wave function of the object is ψ_2; i.e., the object behaves from then on as if I had observed it and had seen no flash. However, even in this case, in which the observation was carried out by someone else, the typical change in the wave function occurred only when some information (the yes or the no of my friend) entered my consciousness. It follows that the quantum description of objects is influenced by impressions entering my consciousness ... ".[9]

Given a passage like this, it is little wonder that many non-physicists, especially those with a penchant for Eastern mysticism, see the flood gates of oneness opening. They should be warned that Wigner's is a minority view among physicists. Most reject the role of consciousness in quantum mechanics. When I think of physicists who invoke consciousness as an "explanation" of the quantum theory, I am reminded of an old joke. A man goes to a psychiatrist to seek help because his brother thinks he is a chicken. "Why don't you show him that he is not?" the psychiatrist asks. "I would," the man says, "except that I need the eggs."

[8]**Wigner, 172 – 3.**
[9]**Wigner, 173.**

Chapter 6

Entanglements

"We continue. That a portion of the knowledge should float in the form of disjunctive conditional statements between the two systems can certainly not happen if we bring up the two from opposite ends of the world and juxtapose them without interaction. For then indeed the two "know" nothing about each other. A measurement of one cannot possibly furnish any grasp of what is to be expected of the other. Any "entanglement of predictions" that takes place can obviously only go back to the fact that the two bodies at some earlier time formed, in a true sense, one system, that is, were interacting and have left behind traces on each other. If two separated bodies, each by itself known maximally, enter a situation in which they influence each other, and separate again, then there occurs regularly that which I have just called entanglement of our knowledge of the two bodies".

Erwin Schrödinger[1]

"To summarize, Bell's theorem showed, in 1984, that either the statistical predictions of quantum theory are false or the principle of local causes is false. In 1972, Clauser and Freedman performed an experiment at Berkeley which validated the relevant statistical predictions of quantum theory. Therefore, according to Bell's theorem, the principle of local causes must be false. The principle of local causes says that what happens in one area does not depend upon variables subject to control of an experimenter in a distant space-like separated area. The simplest way to explain the failure of the principle of local causes is to conclude that what happens in one area

[1]**"The Present Situation in Quantum Mechanics", a translation of Schrodinger's 1931 paper to be found in *Quantum Theory and Measurement*, edited by John Archibald Wheeler and Wojciech Hubert Zurek, Princeton University Press, Princeton, 1983, 161.**

does depend upon variables subject to the control of an experimenter in a distant space-like separated area. If this explanation is correct, then we live in a non-local universe (locality fails) characterized by superluminal (faster than light) connections between apparently "separate parts."

Gary Zukhav[2]

"Bell's theorem supports Bohr's position and proves rigorously that Einstein's view of physical reality as consisting of independent, spatially separated elements is incompatible with the laws of quantum theory. In other words, Bell's theorem demonstrates that the universe is fundamentally interconnected, interdependent, and inseparable. As the Buddhist sage Nagarjuna put it, hundreds of years ago Things derive their being and nature by mutual dependence and are nothing in themselves.

Fritjof Capra[3]

In 1907, Einstein published a paper in the *Annalen der Physik*[4] in which he tied up some loose ends left over from his great 1905 paper on the special theory of relativity—"special" because it did not deal with accelerated motions. I will describe one of the loose ends in more modern language. Suppose that the theory of relativity is valid, but suppose there exist particles that can move faster than light. Such hypothetical particles are now called "tachyons."[5] We must be clear about something. An ordinary particle like an electron, assuming the theory of relativity is correct, can never be accelerated to the speed of light, let alone faster. Tachyons always move at speeds greater than light. It turns out that this does not violate the theory of relativity but it leads to very strange behavior indeed.

Let us suppose that you have a tachyon emitter. At some time 't' you emit a tachyon and at a later time it is absorbed by a tachyon absorber. This is a perfectly causal sequence. Einstein considered how an observer

[2]Gary Zukhav, *The Dancing Wu Li Masters,* Perennial Classics, New York, 2001, 335.

[3]Fritjof Capra, *The Tao of Physics*, Shambhala, Boston, 2000, 313.

[4]*Annalen der Physik,* 23(1907), 371.

[5]This name was introduced by the late Gerald Feinberg, who was a professor at Columbia University. He actually participated in an unsuccessful attempt to find them.

in motion with respect to you would see the same sequence. It turns out that there is a speed less than the speed of light at which this observer would see these two events occur simultaneously. At speeds greater than this, but still less than the speed of light, the time order of the events would be reversed. This observer would see the tachyon absorbed before it was emitted or equivalently, an anti-tachyon would propagate backwards in time. In one of his essays, John Bell used this scenario to create the perfect murder.[6] Suppose I lure the victim to a spot that I call the origin of coordinates at a time that is noon in the rest system. An observer in the moving system—the perpetrator, me—will also call the time noon at the common spatial origin of coordinates. But now, what I do is to fire my tachyon gun at some time before noon in my system so that it arrives at the origin at noon. To me the sequence is perfectly causal. But if I have adjusted my speed correctly then to you—the observer in the resting system—it will appear as if at noon the victim dropped dead, emitting an anti-tachyon which will be absorbed by my gun at a later time. As far as you are concerned there was no murder. Ruling out such strange behavior requires an assumption beyond relativity. This is usually called "locality." In local theories, superluminal communication is not allowed. Most physicists would regard non-local theories as something of an abomination. I will now begin a discussion of how John Bell became associated with the question of locality or non-locality, beginning with some biographical information about Bell.

John Stewart Bell was born to Protestant working-class parents in Belfast on July 28, 1928. He once described his family to me as being "poor but honest." His father, who dropped out of school at age eight, had a variety of jobs including horse trader. There was no tradition whatsoever in the family for any sort of higher education. Anything above elementary school in Great Britain at that time was neither free nor compulsory. It was expected that Bell would leave school at age fourteen and begin to work to help support the family. But his mother encouraged him to continue his education, and somehow enough money was found so that he could enroll in the Belfast Technical High School—basically a trade school where he studied such things as bricklaying and carpentry. But during high school Bell became interested in philosophy and began reading what he described as "big books on Greek philosophy." However he became disillusioned because

[6]John Bell, *Speakable and Unspeakable in Quantum Mechanics*, Cambridge University Press, New York, 2004, 235 – 6.

he thought that all philosophers did was to attempt to refute each other. He did get a glimmering of physics in his high school and decided that while physics did not deal with the larger questions of life and death, it did deal with the laws of nature. This is what Bell decided to pursue. But when he graduated he was both too young and too poor to continue his education.

Bell tried unsuccessfully to look for work until he hit on a job that was almost perfect considering how his career would later evolve. He was hired by a physics instructor at Queen's College, Belfast, named George Karl Emeleus, to set up laboratory experiments for freshman physics students. Emeleus must have seen something exceptional in Bell, because he also suggested physics books for him to read, and he was allowed to listen to the lectures in freshman physics. After a year, this was 1945, Bell was old enough, and had put together enough money, to matriculate. It was at Queen's College that Bell first began learning about quantum mechanics. What he learned outraged him. He understood the mathematics perfectly well, but he did not understand how it was being interpreted. He had flaming arguments with his professor, Robert Sloane. I asked Bell if he thought then that the theory was simply false. "I did not dare to think it was false," Bell replied, "but I knew it was rotten." With his Irish lilt, Bell said "rotten" with a considerable relish.

What bothered Bell then, and bothered him for the rest of his life, were questions like 'what is the Schrödinger wave function?'. The standard interpretation tells us that the wave function provides information about a quantum mechanical system, such as the probability that, when measured, "observables" have some particular value. But is this all we have—"information?" Isn't there something else behind the scenes? Bell used to conjecture that while physicists when they teach, or even use, the quantum theory, seem to accept the idea that all we have is information, in their heart of hearts they really believe that the wave function is not the whole story. The information contained in it refers to something else—something not revealed—a hidden reality. The Wizard of Oz may put on a great show, but behind the screen there is an old man from Omaha turning the cranks.

Bell graduated from Queens with first class honors in physics and decided that he could no longer freeload off his parents. He had been living at home while he was at Queens. He got a job with the Atomic Energy Research establishment in Malvern, Worcestershire. where there was a group designing a linear accelerator—an accelerator, unlike a cyclotron where the particles move in a circle—where the particles move in straight lines. There are some advantages to this when it comes to how the particles radiate. There, he met his future wife Mary, a Scottish physicist who had also come to Malvern. He also grew a red beard to cover up a scar acquired in a motorbike accident. For the rest of their careers both Bells continued to work on the design of accelerators. John had no intention of studying for a higher degree, but in 1953 he was selected for a program of the Atomic Energy Authority which allowed a year's study towards a Ph.D. John decided to go to Birmingham to study with Rudolf—later Sir Rudolf—Peierls who was one of the great 20th century teachers of physics. By this time Bell had published papers in accelerator design, but he had also been thinking about the foundations of quantum theory.

While still at Queens Bell had read Max Born's book, *Natural Philosophy of Cause and Chance.*[7] Born, who died in 1970, was one of the founders of the quantum theory, although his work was not recognized for a Nobel Prize until 1954. He escaped from Germany in 1933. He took up a position in Cambridge and then at the University of Edinburgh. But in 1948, he delivered the so-called Waynflete Lectures at Oxford. These became the basis of his book which was published in 1949. Born presented a masterful survey of all 20th century physics including, of course, quantum theory. Born was at odds with Einstein about the interpretation of the theory. He had proposed its probability interpretation, inspiring Einstein's famous letter to him about God not playing dice. It was a passage in the lectures that struck Bell. By 1949 it had become clear that there were problems in things like quantum electrodynamics. While this theory made fantastically accurate predictions, quantities that should be finite were infinite in the theory. Born addressed the question of whether these difficulties could be cured by some return to an underlying classical physics.

> "I expect that our present theory will be profoundly modified. For it is full of difficulties which I have not mentioned at all—the self energies of particles in interaction and many other quantities, like collision cross-sections, lead to divergent integrals. [A divergent integral becomes infinite.] But I should never expect that these difficulties could be solved by a return to classical concepts. I expect just the opposite, that we shall have to sacrifice some current ideas and use still more abstract methods. However,

[7] **Max Born, *Natural Philosophy of Cause and Chance,* Dover, New York, 1964.**

> these are only opinions. A more concrete contribution to this has been made by J.v. Neumann in his brilliant book *Matematische Grundlagen der Quantenmechanik*. He puts the theory on an axiomatic basis by deriving it from a few postulates of a very plausible and general character, about the properties of "expectation values" (averages) and their representation by mathematical symbols. The result is that the formulation of quantum mechanics is uniquely determined by these axioms; in particular no concealed parameters can be introduced with the help of which the indeterministic description could be transformed into a deterministic one. Hence if a future theory should be deterministic, it cannot be a modification of the present one but must be something essentially different. How this could be possible without sacrificing a whole treasure of well-established results I leave to the determinists to worry about."[8]

Bell was much taken by this and also by the fact that, in early 1952, David Bohm published two papers which did precisely what von Neumann said was impossible. Before I describe Bell's resolution of this paradox and what it led to, let me re-iterate what Bohm did. First, let me say what he did not do. He did not produce a theory that supplants quantum theory. Bohmian mechanics, as it is often called, although never by Bohm, is a re-interpretation of non-relativistic quantum mechanics. I will come back to the non-relativistic issue later. It claims to reproduce all the standard results of the quantum theory in a deterministic way. It is sometimes called a theory of "hidden variables," but this is misleading. There is nothing hidden about the trajectories of the particles in Bohmian mechanics. Each trajectory is explicit and classical. These particles are the primary subjects of the theory. The secondary subjects of the theory are the Schrödinger guide waves that guide them. For the sake of historical fairness it should be mentioned that, after the invention of the quantum theory, de Broglie presented a very similar model. He abandoned it after he was subjected to some scathing and ultimately irrelevant criticism by Pauli.

As I have mentioned earlier there are two basic equations in the theory. There is the Schrödinger equation that determines the guiding wave.

---

[8] **Born, 109.**

Once you use this equation to solve for the wave function, you can use this solution in a very classical-looking equation for the particles that determines their trajectories. Since these trajectories are perfectly classical one may well wonder where in this lies such quantum mechanical necessities as the Heisenberg uncertainty principles. While it is true that any solution of the Bohmian Schrödinger equation determines a unique trajectory of the particles with no uncertainties, the initial conditions that determine the wave function are uncertain, and this reflects itself in the trajectories. This is reminiscent of some of the discussions of determinism that arose after the uncertainty principles were discovered. It used to be said that the Newtonian universe evolved like clockwork. Once the initial conditions were given—the positions and momenta of all the particles—the future was completely determined. But if you could never precisely determine these initial conditions as Heisenberg's uncertainty principles said, then you were liberated from this deterministic straitjacket. I leave it to the philosophers to elaborate on this discussion. In the case of Bohmian mechanics it is the uncertainty in the initial conditions that generates the uncertainties in the trajectories. One might object that this is something that is imposed on the theory and not, as in ordinary quantum mechanics, derived from it. The real question is, is the game worth the candle? Is trying to resolve some unease about the interpretation of the quantum theory worth all the trouble of Bohmian mechanics?

To most working physicists the answer is very probably "no." Just as one can successfully ride a bicycle without delving into the gyroscopic principles that govern turning, one can use the rules of quantum theory to make calculations and predictions without worrying a great deal about what it all means. This has changed to some extent since Bell's work, and many distinguished physicists now feel it is legitimate to look into these matters. Bell was such a good physicist that people felt that if he could do something, so could they Bohmian mechanics is only one avenue. There are others. Take the matter of the double slit which caused me so much bewilderment when I was a college freshman. Electrons, say, impinge on a barrier perforated by two closely spaced slits. There is a detector behind the barrier for the electrons. If the intensity of electrons is reduced so that they enter the apparatus singly, nonetheless over the course of time they build up at the detector, reproducing the wave pattern which a wave would have produced if it had gone through the slits. If one closes one of the slits the pattern changes. The question I asked myself all those years ago was how does the electron "know" that one of the slits has been closed when it passes through the other? Ordinary quantum mechanics does not really answer this question. In fact it dismisses it. There is a rule that tells you how to compute the probability that the electron will land in one spot or the other on the detector, and that is pretty much that. In Bohmian mechanics there is an answer. If both slits are open the guide wave goes through both of them. The interference of the waves reflects itself in the trajectories of the electrons which go through either one slit or the other. There is something satisfying about this, but there is another problem with the whole scheme that was first clearly presented by Bell. I will get to it after I give a bit more of his biography.

When he went to Birmingham for his sabbatical in 1953, Bell had already published some papers on accelerator design. Because of this Peierls told Bell that he would not be treated as a "beginner" but was expected to give a seminar on his work. Bell gave Peierls two choices of subject: accelerator design or the foundations of the quantum theory which would deal with Bohmian mechanics. Peierls made it very clear that he did not want to hear about Bohmian mechanics, for which he had no use. Peierls always maintained that there were no problems to be solved in the interpretation of the theory. Much later he did study and admire Bell's work, although he still took the position that the "problems" had to do with muddled thought. Bell did his Ph.D thesis under the supervision of a then-young assistant to Peierls named Paul Matthews, who went on to have a very distinguished career. After his sabbatical year, Bell went to Harwell, another station of the Atomic Energy Authority. It had a small group devoted to basic research in nuclear physics. He remained there for a few years and then decided that the group did not really fit into anything else the laboratory was doing so he should try to relocate. As it happened, he had already consulted on the design for an accelerator that was being built at CERN in Geneva. He and Mary were offered jobs in 1960, and they spent the rest of their careers at CERN.

Bell had a very strong work ethic. He felt that he was being paid by CERN to do accelerator design and elementary particle theory and not to study the foundations of quantum theory. So that is what he did. But he could not help thinking about the quantum theory. When he was still at Harwell he had a colleague named Franz Mandl who was German-speaking. This was important because there was no English translation of von Neumann's book available then. Once Mandl translated the relevant part of the book Bell saw at once what the problem was with von Neumann. Von Neumann had assumed that the average of sum of quantities is the same as the sum of the averages of them. This is almost trivial for averages in conventional quantum theory would necessarily apply to any deterministic alternative. Bell realized that it did not apply to Bohmian mechanics so that von Neumann's argument was irrelevant. He let the matter rest until 1963, when he and Mary got a sabbatical leave from CERN, part of which they spent at Stanford. He felt that he was now free to work on anything.

Bell first published a review article which he called "On The Problem of Hidden Variables in Quantum Mechanics." This was a summary of all the work he had done on the subject since 1952. Most of the article deals with the question of why various proposed "proofs" did not rule out Bohm-like interpretations of quantum theory. But at the end Bell raises an issue that was only briefly alluded to in Bohm's papers. Suppose you have two Bohmian particles in interaction. This gives rise to two equations of motion, one for each particle. But in general the two particles get intertwined in these equations. The trajectory of one depends on instantaneous knowledge of the behavior of the other even if the two particles are spatially separated. In short, the theory is non-local. It reproduces the results of the quantum theory but at a very heavy cost. He then adds a sentence, "However it must be stressed that, to the present writer's knowledge, there is no proof that any hidden variable account of quantum mechanics must have this extraordinary character."[9] To this is appended a footnote that reads, "Since the completion of this paper such a proof has been found (J.S. Bell, Physics 1, 195 (1965))." This is a reference to the paper in which Bell proved his theorem. Because of this work and what it led to, Bell had been nominated for the Nobel Prize at the time of his death on October 1st, 1990.

To understand what Bell did, we must back up to 1910. This was the year that the New Zealand-born physicist Ernest Rutherford, and his young associates at Manchester, discovered the atomic nucleus. Rutherford then presented the model of the atom which has been with us ever since. Here is its modern version. There is a tiny massive nucleus made up of neutrons and protons which is surrounded, and at a great distance from the nucleus, by a sufficient number of negatively charged electrons so as to balance the charge of the protons. When Niels Bohr came as a postdoctoral fellow from Denmark to Rutherford in 1911 he had available to him this model of the atom as well as the puzzle it posed. What kept the whole thing stable? What prevented the electrons from just collapsing into the nucleus,

---

[9] ***Reviews of Modern Physics*, 38, 447 – 52, (1996).**

considering that they were electrically attracted to the protons? After his return to Denmark, Bohr proposed an answer which he published in 1913.

In his model, the electrons are restricted to certain orbits. In our language, the orbits are "quantized." The orbit with the least energy—the "ground state"—is by hypothesis stable. But electrons in the ground state can be excited to states of higher energy. When they return to the ground state, radiation is emitted as the electrons transition from the states of higher energy to those of lower energy. The energies of this radiation are determined by the energies of the electron in these quantized orbits. Indeed, the difference of energy between one orbit and another is the energy that is radiated. This means that, instead of a chaos of emitted radiation, only certain frequencies are observed. This explains why collections of excited atoms emit the beautiful spectra of light that characterize these atoms like fingerprints.

For the next decade or so after Bohr published his work, there was practically a cottage industry of physicists applying it to more and more complex spectra. All of this—a mixture of classical and new physics—came to be called the "old quantum theory." By the 1920s there were some ineluctable problems which the old quantum theory could not solve. The atomic electron states were characterized by so-called "quantum numbers." In addition to the energies, the electrons in these states carried an orbital angular momentum. This angular momentum was "quantized", meaning that not only did the total angular momentum—unlike classical physics—have a limited number of values it could take on, but for a given value, there were only a limited number of directions to which the angular momentum could point. These added quantum numbers gave rise to additional spectral lines since there were more ways that the electron could make transitions. But there was a troubling doubling of some of the lines that nothing seemed to explain.

Sometimes in physics, when it comes to a conundrum like this it takes young physicists with nothing to lose to resolve it. In this case the two young physicists were both Dutch graduate students; George Uhlenbeck and Samuel Goudsmit. Goudsmit knew a lot about the spectra and Uhlenbeck had a deeper understanding of theoretical physics. In 1925, they proposed that the electron had an additional angular momentum that had nothing to do with its motion. This became known as "spin." In their original picture they saw the electron as a tiny ball of charge spinning like the earth around an axis. But this homey picture is untenable. The electron would have to spin so fast that the exterior would be moving faster than the speed of light. Moreover, to fit the data the electron can only spin pointing in two directions—"up" and "down." Spin is a quantum mechanical property intrinsic to any particle. The introduction of spin resolved the doubling of some of the spectral lines. In fact, although Goudsmit and Uhlenbeck did not realize this, there was data from an entirely different source. There was an experiment that had been done in 1922, that provided conclusive evidence for the existence of spin.

When the German physicist Otto Stern returned from his service in the First World War, he became an assistant to Born in Frankfurt. Born had become interested in what are known as molecular beam experiments. Molecules are heated in a furnace and then allowed to escape into a directed beam. These molecules are electrically neutral so they do not get perturbed by stray electric fields that could overwhelm any subtle effects. Born put Stern onto working with these beams. Stern acquired a collaborator—Walther Gerlach—a very skilled experimental physicist. Stern wanted to test a prediction of the old quantum theory. This had to do with the different directions the angular momentum of the molecule could point. Stern chose the silver molecule because it had one electron outside a closed core of electrons. The total angular momentum of the atom is composed of the angular momentum of the electron and the angular momentum of the core, neither of which were known. But this angular momentum is associated with an atomic magnet whose orientations are those of the total angular momentum. One has in effect a tiny magnet that can point in a limited number of directions, reflecting the quantization of the atomic angular momentum which was unknown. If this tiny magnet interacts with a strong magnetic field which, in turn points in some fixed direction, then depending on how the angular momentum of the atom was oriented, the molecule in the beam would follow one trajectory or another in the external magnetic field. By collecting these atoms on a detector, one could see from these depositions the effect of these different trajectories—how many lines were produced on the detector. The main thing that Stern wanted to test was the quantization of the angular momentum, which Bohr had predicted. If the angular momentum was classical it could point in any direction, so on the detector there would be a smear. In fact, what turned up were two distinct lines. This showed that the angular momentum was quantized, but the fact that there were only two lines was mysterious. It took five years after this experiment was done before this observation was connected to electron spin. The reason was that, until 1927, it was not known that the core of electrons in the silver atom carried no angular momentum so that the entire angular momentum of the atom was that of the exterior electron, which had no orbital angular momentum. Therefore, the whole effect was due to electron spin. There were two possible trajectories in the Stern–Gerlach magnet because the spin of the electron could point only "up" or "down."

I bring all this up because Stern–Gerlach-like magnets play an essential role in Bell's analysis. It all began with Bohm's text on quantum mechanics—*Quantum Theory*—which he published in 1951.[10] Having studied quantum mechanics at that time, I can testify that Bohm's text was the only one I know of that made an extensive study of the interpretation of the theory. One of the things that Bohm presented was his version of the Einstein–Podolsky–Rosen paper. They had considered wave functions that represented entangled momenta which made their arguments rather complicated. Bohm focused on spins which could point only up or down.

In the units in which spins are measured, the electron has spin, as do the neutron and proton. If we put two electrons together then their net spin can be 1 or 0. For his purposes Bohm considered the spin zero state which is known as the "singlet." To get zero net spin, one of the electrons must have spin "up" and the other spin "down", but in the absence of measurement we do not know which is which. In an obvious notation the wave function has the form $\uparrow(1)\downarrow(2)-\downarrow(1)\uparrow(2)$. The minus sign is essential to make the net spin zero. This wave function represents the "entanglement" of the two electrons and indeed, in many ways, it is the paradigmatical example of entanglement.

Suppose there was a hypothetical particle of spin zero that could decay into two electrons. These electrons would be created in the singlet state. If the particle was at rest at the time of the decay, then the electrons would fly off in opposite directions with the same speeds so as to conserve momentum. Unless one of the electrons interacted with something, quantum mechanics tells us that the spin entanglement would endure no matter how far apart the electrons were. Now suppose that I have a Stern–Gerlach magnet somewhere. (The actual Stern–Gerlach experiment was done for technical reasons with silver atoms and not free electrons.) When one of the electrons enters the magnet it can potentially have spin up or down. Which one it has is revealed by detecting its trajectory—where it arrives at the detector. Before this detection, all one can say is that a given electron has the potentiality of having its spin up or down. If we have a collection

[10]**Bohm, 1951.**

of these decaying particles, they will keep supplying electrons to the magnet. Half the time these electrons will be spin up and half the time spin down. If I am doing the detection at one of the magnets the distribution of spin up and spin down will appear entirely random. Suppose I note these different measurements in my lab book with a + and – for spins up and down respectively. Then the entry will look after a while something like ++−+−+−−+++... and so on. Now, you also have a Stern–Gerlach magnet which is farther away from the electron source in the middle than mine, but oriented in the same way that mine is. Your electrons arrive a little later than mine and you also have a lab book in which you note the pluses and minuses. To you they also seem entirely random. Unless we compare notes this is what we will conclude—a random distribution of spins at each magnet. But later on we do compare notes. If we know our quantum mechanics we will not be surprised at the result. The spins measured at our two magnets are perfectly anti-correlated. Whenever I note a plus you will note a minus and vice versa.

If I put myself in the state of mind I was in when I was a college freshman first learning about the two-slit experiment with the single electrons, the question I would ask is how does the distant electron "know" that its mate has just had its spin measured which forces the anti-correlation? In standard quantum theory with no superluminal signals allowed, there is no answer. It is not even a question. The theory tells you how to calculate the correlation and that is that. To console yourself you might think of the following situation. I have a large coin which someone has sliced into two pieces, one with heads and one with tails. This person has put the two halves in sealed bags without telling me or you which is which. I give one of the bags to you and you go off to Kathmandu. There is a fifty-fifty chance that when I open my bag the coin will have the head. I do so and, lo and behold, I do have a head. At that instant, the probability that you have a head drops to zero. There is nothing mysterious about this. Probability reflects the state of knowledge and when this changes so do the probabilities. The difference between this and the quantum mechanical situation is that if I find a head I can be sure that it was always a head. I do not have to do a measurement. But in quantum theory I cannot make that assumption. If I do then I can generate contradictions with the uncertainty principles. This is the sort of thing that troubled Einstein. He called this sort of quantum mechanical correlation a "spooky" action at a distance. Bohmian mechanics does provide an explanation. When the two particles are entangled, whatever one does influences the behavior of the other instantaneously. In the paper in which Bohm discusses this he claims that there is no contradiction with relativity because no "information" is being transmitted this way. Perhaps, but what about causal sequences as viewed from different reference frames? It seems to me that once you open the Pandora's box of this kind of instantaneous, non-local influence sharing, no one can be sure what is going to pop out.

Bell understood all this. The question he asked himself was, can you have the quantum mechanical correlations generated by some classical hidden variable theory, unlike Bohm's, with only local interactions? In thinking about this Bell proposed a variant of the two parallel Stern–Gerlach magnets that no one had thought of. Suppose you rotate one or both of the magnets, what happens to the correlation? Quantum mechanics has a definite answer. If the two magnets are parallel then there is the perfect anti-correlation we just discussed. If they are perpendicular there is no correlation at all. In between, when the magnets are rotated at an angle a, the correlation is given by $-\cos(a)$ which for $a = 0^\circ$ and $a = 90^\circ$ gives you the results I just mentioned. The question then became whether any local hidden variable theory could reproduce this correlation. What Bell meant by "local" was that the value of the hidden variables at one magnet did not in any way depend on what was happening at the other magnet. Bell found it easy to exhibit a toy model with these properties for the cases of zero and ninety degrees. But for the angles in between no such model worked.

Bell's general argument is rather abstract and sophisticated. I want to present an analysis which is neither. I don't claim any rigor for what follows. I can tell you that it was suggested by some remarks of Bell, although without the doodads. I am going to imagine I have available to me some incredibly clever automata which I will call "Einstein robots"—the doodads. I can program the robots to do anything except send signals to each other at superluminal speeds. The question is can I program them to reproduce the quantum correlation without quantum mechanics? I present one of the robots with a batch of diatomic molecules, each of whose atoms has spin. I instruct the robot to split the molecule and fling the components in opposite directions. I tell it to do this repeatedly, making sure that the spin ups and downs are opposite in each case, but without prejudice as to which of the components has which spin orientation. You may ask, where is the entangled wave function? There isn't one, which is just the point. I am doing quantum mechanics without doing quantum mechanics. To each of the components I attach an Einstein robot. If the Stern–Gerlach magnets are parallel the robots are instructed to guide the trajectories of the particles suitably according to whether the spin is up or down. This will reproduce the anti-correlation. Now I rotate one of the magnets through a small angle 'a'. I instruct the robots when they encounter the rotated angle to adjust the trajectories so as to reproduce the small angle correlation $-1 + a^2/2$.[11] So far so good. I return this magnet to its original position and rotate the other magnet through the angle -a. With the same instructions to the robots I get the same correlation. But now I pull a dirty trick. I rotate one magnet through a and the other through -a. The poor robots, which are not allowed to communicate, do not know how to do anything but produce the sum of the correlations for a and −a which is $-1 + a^2$. But the quantum mechanical correlation involves the angle 2a which is the angular separation of the two magnets. This correlation is $-1 + 1/2(2a)^2$. If we take, say $a = 1/2$ then the "classical" answer is $-3/4$ while the quantum answer is $-1/2$. This inequality is a very special case of what is known as Bell's inequalities. What Bell had shown is that no local hidden variable theory can reproduce all the results of quantum mechanics. Bohmian mechanics can reproduce them because it is not local. It solves the problem of an explanation in the way that Einstein would probably have liked least. It replaces the "spooky actions at a distance" of quantum mechanics by a theory that has superluminal influences.

It is sometimes said that Bell proved that Bohr was right and that Einstein was wrong. Bell even said as much, with some regret, since he felt that Einstein was asking the natural scientific question and Bohr was replying by saying that the question should not be asked. Bell found much of Bohr's writing on the quantum theory to be very obscure which irritated him to no end. But it is not clear to me that this result would have displeased Einstein. What we do know is that he did not think much of Bohmian mechanics. He wrote a letter to Born in May of 1952 in which he said, "Have you noticed that Bohm believes (as de Broglie did, by the way, twenty five years ago) that he is able to determine quantum theory in deterministic terms? That way seems too cheap to me. But you, of course, can judge this better than I."[12] I have never understood what Einstein meant by "too cheap" nor exactly what he was trying to do, but I am not sure that Bell's results would have made much difference to him.

Having answered the question he set out to answer, Bell had to decide where to publish it. The natural place would have been one of our trade

[11]**For small angles, $\cos(a)=1-a^2/2$.**

[12]***The Born-Einstein Letters,* Walker, New York, 1971, 192.**

journals such as *The Physical Review*. But at the time these all had substantial page charges which were paid by whatever entity, or individual, had submitted the paper, usually out of government grant money. Bell had no intention of paying these page charges himself, nor did he think it was right to have Stanford pay them since he was a guest. However, it turned out that a new journal named *Physics* was actually paying for articles. This is where Bell sent his, which was entitled "On the Einstein Podolsky Rosen Paradox." It was published in 1965 in the first issue.[13] It turned out that the journal, which rather soon went out of business, did not supply free reprints so Bell used up most of his honorarium from the journal to buy reprints. He might have saved his money, since for the next five years he had almost no requests. There was essentially no interest in what he had done and he went back to doing elementary particle theory and designing accelerators. Then he received a letter from a young physicist at Berkeley named John Clauser that changed everything.

Clauser had taken his degree at Columbia University in experimental physics. Columbia had a requirement that a graduate student get at least a B in an advanced quantum mechanics course. Clauser kept getting C's because he could not understand what the quantum manipulations meant. This led him to read Bohm's papers and ultimately, to Bell. There was no one at Columbia who had any interest in these things with whom he could talk to. He finally passed the course and got his degree and a job at Berkeley working in the laboratory of the Nobelist Charles Townes. By this time Clauser had seen how to test Bell's inequalities experimentally, and Townes was sufficiently interested to allow him to use the facilities of the laboratory and to assign him, part time, the assistance of another laboratory physicist, Stuart Freedman. Clauser's important insight was not to use electrons in the experiment—the fact that they were electrically charged would overwhelm any subtle effects—but to use light.

Light quanta—photons—have spin. In this case, spin one. Hence, if you produce a pair of entangled photons you could, in principle, measure the correlations of their spins. Clauser derived a new version of the Bell inequality that applied to this case and which also took into account some of the limitations of an actual experiment. Unknown to Clauser, on the East coast a group of physicists had done the same thing. They were led by the philosopher-physicist Abner Shimony of Boston University, his graduate student, Michael Horne, and a Harvard physicist named

[13] *Physics* 1, **195 – 200.**

Richard Holt. The East and West coast physicists decided to collaborate and they published a joint paper in 1969.[14] It has become a standard reference in the field. In the meanwhile, using these ideas, Clauser and Freedman were preparing their experiment. The first problem was where do you get correlated photon pairs? One idea was to use the annihilation of the electron with its anti-particle, the positron. This does produce a correlated photon pair but positrons are hard to come by in quantity, so they settled on an atomic process using cadmium. By 1972, they were able to announce their results which agreed with quantum mechanics. Then, the experiments were taken up by a French physicist named Alain Aspect. His very high precision results agree with the quantum theory. To most physicists this settles in the negative the question of whether a local, hidden variable theory can reproduce the predictions of the quantum theory.

Bell was, of course, following these developments with great interest, but so were a number of other people, many of whom were not physicists. Bell could easily have become a guru the way Bohm did. But this was not in his nature so he watched all of this with what I would describe as stoic amusement. He gave many popular lectures. He found a device for leading people into the subject—identical twins who had been separated at birth. Bell discovered that there were a number of pairs that had been re-united later in life only to find that they had a remarkable number of things in common, including smoking the same brand of cigarettes. We have an explanation for this—they share the same genes. But what is the explanation for the correlations of the widely separated entangled photons or electrons? If you accept the usual interpretation of the quantum theory, there isn't one.

By now, Bell's theorem seems to have gone beyond physics. From time to time I check some of the tens of thousands of websites of every description on which the theorem is discussed. It is a very mixed bag. Here are two examples. There is a charming site called Kids.Net.Au.[15] One sentence reads, "In 1964 [Bell] demonstrated that quantum mechanics requires superaluminal [?] signaling..." The question mark is in the original. Then there is a site that is called "Shamanic Healing: Why it Works."[16] Here are a couple of quotes:

---

[14] ***Physical Review Letters*, 23, 880 – 84 (1969).**
[15] **http://encyclopedia.kids.ner.au/page/John_Stewart_Bell**
[16] **http://beliefnet.com/story/136/story_13663_2.html**

"Strained by the conflicts between Einstein and Bohr over the ultimate meaning of quantum mechanics, subjected to further stress in Bell's theorem, and finally ripped through in recent tests, the whole cloth of the materialistic picture of reality must now be rejected..."

And "...this means that shamanism finally has an explanation based in modern physics. Shamans can effect change in local reality through spirit helpers working at the quantum level. This is achieved through their ritual action, in which the shaman's consciousness intently focused on a singular objective. For example, 'Take this cancer out of this sick person."

I once asked Bell what he thought the problem with quantum theory was. He laughed and said that if he knew, he might make some progress towards solving it.

Chapter 7

Anyway What The #$*! Do We Know?

"It's only in conscious
Experience that it seems that
We move forward in time.
In quantum theory
You can also
Go backwards in time."

Dr. Stuart Hameroff[1]

"Quantum mechanics allows for the intangible phenomenon of freedom to be woven into human nature."

Dr. Jeffrey Satinover[2]

"But it had also been Schrödinger who had convincingly argued that because the wave function is stubbornly smeared out over configuration space, the abstract space of all possible configurations of particles, until the precise moment of its collapse, it therefore resists all attempts to connect up with a world recognizably like our own. Faced with this intractability, formally known as the "measurement problem", many of the luminaries of physics, from Bohr and Heisenberg on down, took a radical step of denying the existence of an independently existing world altogether and, surprisingly, got away with it.

---

[1]**Stuart Hameroff, *The Little Book of Bleeps,* Captured Light Distribution, 2004. This book, perhaps suggested by the quantum theory uncertainty principles, does not offer page numbers.**

[2]**Jeffrey Satinover, *The Little Book of Bleeps.***

In other, i.e. nonscientific contexts, the difference between those who are committed to an independently existing reality and those who are not is roughly correlated with the distinction between the sane and the psychotic."

Rebecca Goldstein[3]

"When he first met Desplechin in 1982, Djerzinski was finishing his doctoral thesis at the University of Orsay. As part of his studies he took part in Alain Aspect's groundbreaking experiments, which showed that the behavior of photons emitted in succession from a single calcium atom was inseparable from the others. Michel was the youngest researcher on the team."

Michel Houllebecq, *The Elementary Particles*[4]

"Where there's smoke, there's smoke."

John Wheeler

I went through a chess playing phase. I was never good enough so that it became an obsession, but I did study and take a few lessons. I played chess hustlers—and still do—in Washington Square Park in Greenwich Village in New York. The first time I met Stanley Kubrick I discovered that he had been, somewhat before my time, one of those chess hustlers. This led to a match of 25 games—which he won—while he was filming 2001. My account of this in The New Yorker led to a commission from Playboy magazine to cover the Bobby Fischer-Boris Spassky match in Iceland. I tell you this because of a missed opportunity. One time when I was visiting my parents in Rochester I learned that the International Grandmaster Samuel Reshevsky was giving an exhibition. Reshevsky had dominated American chess for several decades and was one of my heroes. He was going to play a simultaneous match with some forty players, each one paying a few dollars for the privilege. I signed up.

In an exhibition like this, the boards are set up on the periphery and the master makes the rounds from the inside. Reshevsky stopped at each board for what seemed like seconds before he made his move and went on to the next player. This gave one some time to contemplate one's next move and to study the other participants. I noticed a few boards from mine

[3]Rebecca Goldstein, *The Properties of Light*, Houghton Mifflin, Boston, 2001, 37–8.

[4]Translated from the French *Les particules é lé mentaires* by Frank Wynne. The English edition is published by Knopf, New York, 200 and this quote is on page 103. The French edition was published by Flammarion, Paris in 1998.

a man with a very distinctive angular face. I did not know who he was but he was certainly not from Rochester. Later when I asked, I was told that it was Marcel Duchamp. By 1923, he had essentially given up art for chess, and in 1925, he came close to winning the French championship. He designed and built most of his own set carved from wood, although a local craftsman did the knights. His wife at the time was not amused and glued the pieces to the board. The marriage did not survive much longer. The missed opportunity? I did not realize that Duchamp had read both Einstein and Henri Poincaré on relativity. How much this influenced a painting like *Nude Descending a Staircase* is much discussed. I could have asked him.

While there does not seem to be a consensus about the influence of relativity on Duchamp there is no doubt about its influence on Lawrence Durrell. In an interview with Dieter Zimmer he said, "The first three parts of my [Alexandria] Quartet develop the three dimensions of space, each from a different point of view or dimension so that they come in conflict with each other. The fourth then adds time to the dimensions of space. You see there is a parallel to the theory of relativity...".

It is said that quantum mechanics played a role in Durrell's *The Avignon Quintet.* I don't see it, but I may have missed something. It is also said that quantum mechanics plays a role in Thomas Pynchon's novel *Gravity's Rainbow.* Pynchon did briefly study engineering physics at Cornell. There is real science in *Gravity's Rainbow.* Here is an example. One of Pynchon's characters is called Roger Mexico. He is a statistician. He decides to apply statistical theory to the explosions of the V2 rockets fired by the Germans towards London. He divides London into squares and plots the number of explosions in each square. He discovers that they follow what is known as a Poisson distribution which means that the landings are by chance. This means that whatever the guidance system the Germans had on the rocket, while it was good enough to target London, it was not good enough to target, say, Buckingham Palace, a useful piece of information. Remarkably, there was a British actuary named R.D. Clarke who did just this. He published a one-page paper on it in 1946. Pynchon must have seen it, because Clarke and Mexico use the same mathematical notation. This is not quantum mechanics but it is fascinating nonetheless. There are two novelists who do use quantum mechanics in their novels, Rebecca Goldstein and Michel Houellebecq. First, Goldstein.

She was born in 1950 into an orthodox Jewish family and grew up in White Plains, New York. She did her undergraduate work at Barnard College and did a Ph.D in philosophy at Princeton. She then taught in various universities and won a MacArthur grant in recognition of her writing. Her teaching has given her a great deal of opportunity to study academic life which, in my experience, is often governed by the partial obverse of Lord Acton's maxim; i.e. the absolute lack of power corrupts absolutely. Tenured faculty, who have come to despise each other, only have the power to make each others' lives as miserable as possible. In her novel *Properties of Light,* in which quantum theory plays an essential role, these characteristics are in full flower.

Much of the action in the novel centers around the physics department of a prestigious eastern university. Princeton? The department chairman is an odious man named Dietrich Spencer and one of the most senior physicists is a man named Samuel Mallach. In an afterword Goldstein tells us that the character of Mallach was inspired by David Bohm, whom she knew. She has changed many of the details. Mallach/Bohm has a beautiful daughter named Dana while Bohm was childless. Mallach/Bohm's wife was an alcoholic while Bohm's wife, Sarah, wasn't. Mallach/Bohm's father sold used furniture in Scranton, Pennsylvania while Bohm's father sold used furniture in Wilkes-Barre. Both fathers were devoutly Jewish and neither son was. Of much greater significance, both men wrote papers on a hidden variable deterministic interpretation of the quantum theory. This has caused Mallach/Bohm to be exiled in his physics department which dismisses his work as useless, or worse. He has been assigned to teach in perpetuity a course for non-physicists called "Physics for Poets". In my experience, there are not many poets who take this course if they can avoid it. Mallach/Bohm has come to love the poetry but he is an abominable teacher whom his students despise. He, in turn, despises—hates—Spencer and only comes to the department when he has to teach his course. He has completely abandoned physics and can hardly remember the content of his own papers.

Into this witches' cauldron is injected Justin Childs. He is an extremely brilliant young theoretical physicist who has learned his quantum mechanics at Paradise Tech in California, He has no problem at all with the formal aspects of the theory. It is its meaning that he has trouble with. Then he comes upon the paper of Mallach/Bohm. Goldstein writes,

> "Justin had read Mallach's paper with something like amazement, for in it he saw clearly that the impossible had been done: an objective model for quantum mechanics. Mallach had formulated a hidden-variable version of quantum physics that had accomplished wonders for the

> material world, saved it from the mathemysticians, the kabbalists of Copenhagen, waving their hands and intoning their obfuscations, turning matter into a miserable ghost of itself, suspended in the not-quite-here-not-quite -there of quantum paradox, in which mess they reveled. Mallach's work, Mallachian mechanics as Justin would eventually dub it (though not without demurs from Mallach himself), was the very countercharm to break the vicious spell...".[5]

I was reminded of John Bell's discovery of Bohm at about the same point in his career. He too had been dissatisfied with the "kabbalists of Copenhagen" and had learned that the great von Neumann had "proven" that a hidden variables theory could not reproduce quantum mechanics. Then in 1952 came the papers of Bohm. Bell writes,

> "But in 1952 I saw the impossible done. It was in papers by David Bohm. Bohm showed explicitly how parameters could indeed be introduced, into nonrelativistic wave mechanics, with the help of which the indeterministic description could be transformed into a deterministic one. More importantly, in my opinion, the subjectivity of the orthodox version, the necessary reference to the 'observer,' could be eliminated."[6]

As I have mentioned, by 1952, Bohm had left the United States for Brazil. Investigations were weighing down on him because of his communist background so Bell could not have consulted him then if he had wanted to. On the other hand Mallach/ Bohm is in a nearby office so Justin Childs has no problem finding him. At first Mallach/Bohm thinks that Justin is a student. No member of his department comes to consult him about anything. When he realizes that Childs is a faculty member he says that he would prefer that they go to his house. When he gets there Mallach/Bohm takes him to his study and has to find an additional chair. There then ensues an odd monologue in which Mallach/Bohm says things like " Perhaps...what we learn from the wave function is nothing of the system but only something of the systematizer, just as dream descriptions are only

[5]**Goldstein, 41–2.**
[6]**John Bell, 160.**

revelatory of the dreamer."[7] This goes on for awhile when Mallach/Bohm interjects, "I do not know what it is you want from me. I cannot fathom what it is you are after. Why don't you go speak to my daughter? I have a daughter, you know."[8] Enter the beautiful Dana.

Dana is attracted to Childs and seduces him on the spot. It does not seem to surprise, or even much interest, Mallach/Bohm that Childs appears at breakfast the next morning. Much of the remainder of the book is devoted to this entangled and ultimately tragic love affair. One of the things that holds it together is that the three of them embark on a project to fill in the lacunae in Mallach/Bohm's paper. Each of them brings a different skill set to the task. Mallach/Bohm brings his original insights. Childs brings his prodigious mathematical skill and Dana, while she has less formal training than they do, has a phenomenal instinct for the physics.

A reader of the novel does not really need to know quantum theory. In fact one might be better off if one does not understand the issues and simply takes the theory as metaphor. I am, fortunately, or unfortunately, not in this position. While I was reading what she writes about their project I kept thinking of something that Wolfgang Pauli once said in a similar situation. I offer it to you *toute proportion gardée.* "I know much," he said, "I know too much. I am a quantum elder." Here is one of her explanations,

> "Mallachian mechanics seems, at first blush, more irreconcilable with relativity theory than other formulations of quantum mechanics, as the very few physicists who had bothered to take notice of his work emphasized. But this is only because it brings into the light what the other formulations obscure, which is the utterly startling, but nonetheless utterly undeniable fact that nature is, in the word of the physicist "nonlocal": events can have instantaneous influences on other, far-distant events. A crude way of seeing that non-locality—and therefore quantum mechanics—is at odds with Einstein's relativity theory is to remember that it is fundamental to relativity that nothing can travel faster than light, whereas these instantaneous propagations of influence seem to indicate superluminal, even infinite velocities. The cognitive dissonance can be expressed

[7] **Goldstein, 49.**
[8] **Goldstein, 53.**

> in more subtle and technical terms, involving the relativistic conditions of Lorentz invariance, which, when combined with quantum nonlocality, allows for such unacceptable anomalies as "backwards causation," which would be the future affecting the past."[9]

The first thing that struck me about this remarkable paragraph was the assertion that relativity and quantum theory are "at odds" with each other. Even Einstein never made a claim like this. By using the term "at odds" I can only assume that the implication is that both theories cannot be simultaneously correct. If this should ever turn out to be true it would be one of the most important discoveries in the history of physics. So far, every experiment done shows that both theories are correct. But more fundamentally, Goldstein appears to believe that Bohm's theory is at odds with relativity because the "influences" travel at superluminal speeds. Bohm, and the other physicists who work on this, are insistent that nothing in this theory that carries information can move faster than light. Whether one can make a relativistic version of the theory is a subject of active investigation. Later Goldstein writes that in quantum mechanics, "events can have instantaneous influences on other, far-distant events." What is an "event" and what is an "influence"? Bell was fond of the following little example. At the death of Queen Elizabeth, Prince Charles instantaneously becomes King. Surely this in an "event" having an "instantaneous influence" on another, perhaps far off event. No one would argue that there is anything especially odd about this, nor that it violates the theory of relativity. If Prince Charles were to learn instantaneously of the death of his mother that would be another matter. I think that one can make the case that the quantum mechanical situation is something like this. Probabilites change instantaneously but we cannot communicate this with signals that travel faster than the speed of light.

Let us return to the double Stern–Gerlach experiment. So long as the singlet state of the two spinning electrons is not interfered with, the spins remain entangled. This reflects itself in the correlation of the two spins. No matter what the orientation of the magnets is each observer will observe a random pattern of spin ups and spin downs. Until the observers compare notes they have no evidence of the correlation. Once they compare notes they see that the correlations changed when one of the magnets was rotated

---

[9]**Goldstein, 71–2.**

even when the electrons were in flight. There is no contradiction with relativity or anything else except common sense. Given our experience with most other things, it is natural to demand an "explanation" for this bizarre behavior. It is in these explanations where the troubles begin—where one begins to introduce the "instantaneous propagation of influences." If we are willing simply to accept the quantum mechanical prediction without seeking an "explanation" then these issues do not arise.

Bohmian–Mallachian mechanics offers an explanation but at a cost. There are instantaneous influences. But do these "influences" violate the theory of relativity? On this matter Bohm was categorical. In the second of these two papers he wrote, "The reason why no contradictions with relativity arise in our interpretation despite the instantaneous transmission of momentum between particles is that no signal can be carried this way."[10] He does not tell us what a "signal" is nor does he tell us whether there is a problem with causality. Of course we now know, thanks to Bell, that these hidden variable theories, if they are to produce results in agreement with quantum mechanics, must be "non-local." In his argument for his inequality Bell makes it very clear what he means by non-local. There is a hidden variable governing what happens at each magnet but the values at one magnet do not depend in any way on the other. This work of Bell makes whatever task the three characters in Goldstein's novel are pursuing much more difficult. In fact, I cannot tell from the novel what they are trying to do. But I am pleased to report that, whatever it was, they succeeded and that Dana submitted a paper with Childs as the first author. As the novel ends, she is awaiting the referee reports.

Houellebecq is quite something else. It is not possible to imagine Goldstein's novel without the quantum theory, but it is perfectly possible to imagine Houellebecq's novels without it. Houellebecq introduces the quantum theory because it interests him. Where does this scientific interest come from? Michel Houellebecq was born Michel Thomas on the island of Saint Pierre de La Ré union on the 26th of February, 1956. His mother Janine Ceccaldi Thomas was a medical doctor and something of a free spirit. She married René Thomas in 1953 but the marriage broke up when she had an affair with another man and gave birth to a daughter. Michel had been sent to live with his maternal grandparents but in 1961 his father re-installed him with his own mother whose maiden name had been Houellebecq, a name that Michel Thomas adopted when he began his

---

[10]**Wheeler, Zurek, 390.**

literary career. He also altered his birth date to 1958. Young Thomas was a precocious child—always first in his class. He was a natural to enter one of the Grandes Écoles. He chose to orient himself towards biology which did not exempt him from two years of preparation for the entrance examinations. This preparation, which was intense, involved physics and, above all, mathematics. In the event he chose to enter one of the less prestigious of the Grandes Écoles—the Institut Nationale Agronomique—which trained people to work in agriculture. As a specialty, Thomas chose ecology. He graduated in 1978, and for the next three years, despite submitting his resumé innumerable times, he could not find a job so he entered a film school where he remained for two years. He then began a career as a computer systems analyst, finally at the French National Assembly. In the meanwhile he was writing both poetry and his first novel, really a novella, entitled *Extension du Domaine de la Lutte* which deals with a computer scientist. It has hints of the graphic sex that is part of all of his novels and also of a mordant view of human nature. It was based in part on his fellow workers who had no idea, until it was pointed out to them, that the author Michel Houellebecq was their former colleague Michel Thomas. It was his second novel, *Les particules é lé mentaires*, which he published in 1998, that convinced the French literary establishment that they had a supernova on their hands.

Les particules é lé mentaires is about two step-brothers Michel Djerzinski and Bruno Ceccaldi. Houellebecq gives their mother the name Janine Ceccaldi, the name of his own estranged mother. Michel begins his scientific career as a physicist. Houellebecq could have made him any kind of a physicist, since he soon leaves physics for biology, but his choice is very interesting. Michel is the youngest member of Alain Aspect's group at Orsay which is in the process of testing Bell's inequalities. Incidentally, Aspect told me that Houellebecq never consulted him. But Houellebecq gives an excellent account of the significance of the experiments,

> "Aspect's experiments—precise, rigorous, and perfectly documented—were to have profound repercussions in the scientific community. The results, it was acknowledged, were the first clear-cut refutation of Einstein, Podolsky and Rosen's objections when they claimed in 1935 that 'quantum theory is incomplete.' Here was a clear violation of Bell's inequalities—derived from Einstein's hypotheses—since the results tallied perfectly with quantum predictions. This meant that only two hypotheses were possible.

> Either the hidden properties which governed the behavior of subatomic particles were nonlocal—meaning they could instantaneously influence one another at an arbitrary distance—or else the very notion of particles having intrinsic properties in the absence of observation had to be abandoned. The latter opened up a deep ontological void—unless one added a radical positivism and contented oneself with developing a mathematical formalism which predicted the observable and gave up on any idea of an underlying reality. Naturally, it was the last option which won over the majority of researchers."[11]

From my own experience I would say that the "majority of researchers" prefer not to think about the problem at all.

It is clear from his latest novel, *la possibilité d'une î le*, which was published in 2005, that Houellebecq's interest in quantum theory has persisted. In this novel, which is a kind of science fiction, quantum theory is even more irrelevant. Houellebecq simply inserts his reflections on the matter when it suits him. What has caught his fancy is a development by Murray Gell-Mann and his former student James Hartle. Hartle is a frequent collaborator of Stephen Hawking's. Gell-Mann won the 1969 Nobel Prize in Physics for his work on elementary particles. In 1994, Gell-Mann published a popular book on quantum theory, *The Quark and the Jaguar.*[12] It is clear from what he writes that Houellebecq has read this book. Gell-Mann and Hartle, who are old friends, told me that they were surprised to find their work discussed in this novel and had never had any contact with Houellebecq.

The development that led to the work of Gell-Mann and Hartle began with a paper of Dirac's which was published in 1933. It was pretty much ignored until Feynman took up Dirac's ideas for a thesis that he did under the supervision of John Wheeler. It was finished in 1942. Feynman then went off to Los Alamos so he did not publish anything about it until 1946. The essential idea is this. Suppose at some time t, within the limits of the uncertainty principles, measurements have been performed on such things as the position and momentum, and perhaps the spin, of some particle.

[11] *The Elementary Particles,* **103.**

[12] **Murray Gell-Mann,** *The Quark and the Jaguar,* **W.H, Freeman, New York, 1994.**

The particle then evolves in time while undergoing interactions with various forces. At a later time these same quantities are measured. If this were classical physics we could predict with certainty the results of these measurements. But in quantum mechanics all we can do is to predict probable values. In fact, in the usual interpretation of the theory, none of these properties have any value until a measurement is made. In classical physics such a particle would trace out an orbit and we could be certain that there was such an orbit even though we did not choose to observe it at every point. In the usual interpretation of the quantum theory, we have no right to assume anything about orbits in the absence of measurement. This means that the particle could have had several possible "paths" or "histories" between the times that we did our observations. In computing the probabilities we have to sum over all these histories. This being quantum mechanics these histories can interfere with each other the way that waves do. But interactions can cause these histories to "decohere." Indeed under the right circumstances these decoherent histories behave at least approximately like classical orbits. This is how, according to this picture, the classical world emerges from the quantum theory.

Some adaptive systems—such as us—are capable of observing. Gell-Mann and Hartle call such systems "IGUS"—Information Gathering and Utilising Systems. As such a system evolves it confronts different possible branch histories but follows only one. The others it discards. In his book Gell-Mann writes,

> "An observation in this context means a kind of pruning of the tree of branching histories. At a particular branching, only one of the branches is preserved (more precisely, on each branch only that branch is preserved!) The branches that are pruned are thrown away, along with all the parts of the trees that grow out of the branches that are pruned."[13] Wave functions don't collapse. The discarded bits belong to another history.

All of this seems to have appealed to Houellebecq's poetic imagination. He writes in his novel, *la possibilité d'une î sle* (the translation is mine),

[13] **Gell-Mann, 155.**

> "For an IGUS observer be it natural or artificial, only one branch of the universe can be given a real existence; if this conclusion does not at all exclude the possibility of other branches of the universe, it forbids any access from them to the given observer: in the formulation, at once mysterious and synthetic, of Gell-Mann 'on each branch only that branch is preserved.' The presence even of one community of observers reduced to two IGUS, constitutes the proof of the existence of a reality."[14]

While Houellebecq finds poetry in quantum mechanics others have found spirituality. Here a word of caution is in order. Schrödinger was a student of oriental religions and wrote about them. But this is what he said in the introduction to his book *My View of the World*,

> "There is one complaint which I shall not escape. Not a word is said here of acausality, wave mechanics, indeterminancy relations, complementarity, an expanding universe, continuous creation, etc. Why doesn't he talk about what he knows instead of trespassing on the professional philosopher's preserves? Ne sutor supra crepidam. [Cobbler stick to your last.] On this I can cheerfully justify myself: because I do not think that these things have as much connection as is currently supposed with a philosophical view of the world."

I am going to finish this chapter, and the book, by discussing people who feel that "these things" have everything to do with a philosophical view of the world.

As far as I can tell, the present vogue began when the work of Bell became widely known. For example in 1975, the physicist Henry Stapp wrote that "Bell's theorem is the most profound discovery of science."[15] One might have thought that he would have put the discovery of quantum mechanics itself somewhat higher on the list. But Stapp's evaluation is widely quoted by people who use Bell's work as an argument for various mystical beliefs. Among the first was Gary Zukav who in 1979 published *The Dancing Wu Li Masters.*[16] In the beginning of his book, Zukav, who

[14] ***La Possibilite*, 345.**
[15] ***Il Nuovo Cimento* 29B, 271 (1975).**
[16] **Perennial Classics, New York, 2001.**

is not a physicist, informs us that he got his inspiration for the book when he attended a conference at the Esalen Institute at Big Sur in California. Esalen is better known for teaching New Age mysticism than physics, but if the two can be mixed so much the better. Inspired by his visit, Zukav began talking to several physicists including the aforementioned Henry Stapp. The result of all of this is a popular presentation of such things as elementary particle physics as they were in the mid-1970s. The problem is that thirty years have gone by and in this field that is a millennium. There is a new generation of physicists produced roughly every ten years. This means that physics is now being done by the graduate students of the graduate students of the physicists Zukav interviewed. Even though I have the 2001 edition of his book, the most up to date reference is 1977! Reading this book is like being in a time warp. Indeed, that is the risk of trying to link some set of spiritual beliefs to a contemporary science. It is like building a house on a block of ice. The Dalai Lama, whose interest in science is very contemporary, has said on many occasions—I heard him say it myself—that if there was a conflict between a Buddhist tenet and a proven scientific fact, the Buddhist tenet would have to be modified. What is one to make of a statement like the following—it is typical—in Zukav,

> "Subatomic particles forever partake of this unceasing dance of annihilation and creation. In fact, subatomic particles are this unceasing dance of annihilation and creation. This twentieth century discovery, with all its psychedelic implications, is not a new concept. In fact, it is very similar to the way that much of the earth's populations, including the Hindus and the Buddhists, view their reality.
> Hindu mythology is virtually a large-scale projection into the psychological realm of microscopic scientific discoveries. Hindu deities such as Shiva and Vishnu continually dance the creation and destruction of universes while the Buddhist image of the wheel of life symbolizes the unending process of birth, death, and rebirth which is part of the world of form, which is emptiness, which is form."[17]

I have a test for phrases like "which is part of the world of form, which is emptiness, which is form." I negate the propositions—"which is not part of the world of form, which is not emptiness, which is not form." If I cannot

[17]**Zukhav, 240–1.**

attach more sense to one as opposed to the other, I go on to something else. In this respect, what is "the psychological realm of microscopic scientific discoveries"? Try as I might, I cannot make much sense out of Zukav's book. I also cannot make much sense out of Fritjof Capra's book *The Tao of Physics* which he published in 1975.[18]

Capra has a Ph.D in physics. The theory of elementary particles he worked in thirty or more years ago had at the time a great vogue. Now no one has much interest in it, although you would not be able to tell this from reading his book. On the other hand, he has kept up to a certain extent with developments, and the latest edition of his book published in 1999 reflects this. In an afterword to the most recent edition Capra says that he is pleased that no physicist has found anything wrong in his presentation. Pauli had a category of things that were not even wrong. They were not coherent enough to be able to assign truth or falsity to them. Here is a fairly typical paragraph in Capra:

> "The conception of physical things and phenomena as transient manifestations of an underlying fundamental entity is not only a basic element of quantum field theory, but also a basic element of the Eastern world view. Like Einstein, the Eastern mystics consider this underlying reality as the only reality; all its phenomenal manifestations are seen as transitory and illusory. This reality of the Eastern mystic cannot be identified with the quantum field of the physicist because it is seen as the essence of all phenomena in this world, and consequently, is beyond all concepts and ideas. The quantum field on the other hand, is a well-defined concept which only accounts for some of the physical phenomena. Nevertheless, the intuition behind the physicist's interpretation of the subatomic world, in terms of the quantum field, is closely paralleled by that of the Eastern mystic who interprets his or her experience of the world in terms of an underlying reality. Subsequent to the emergence of the field concept, physicists have attempted to unify the various fields into a single fundamental field which would incorporate all physical phenomena. Einstein, in particular, spent the last years

[18] **Shambhala, Boston, 2000.**

> of his life searching for such a unified field. The Brahman of the Hindus, like Dharmkaya of the Buddhists, and the Tao of the Taoists, can be seen, perhaps, as the ultimate unified field from which spring not only the phenomena studied in physics, but all other phenomena as well."[19]

I like especially the "perhaps" in the last sentence. Capra's book has had many readers who admire it. I am not one of them.

The *ne plus ultra* of this vogue is the film *What the #$*! Do We Know?* It is a docurama that combines scenes with actors, animation and interviews. When I first saw the film, and especially the interviews, I had a sense of *déjà vu*. In the 1960s I spent a good deal of time in a coffee house restaurant in Greenwich Village called the Café Figaro. The food was good, the prices reasonable and they allowed you to play chess. You could go to any table in the Figaro and find people who talked and looked like the ones in the film. I wonder if some of them were the same people. At any table, including I am sure my own, the most amiable nonsense was being propounded with the greatest sincerity although I do not remember the word "quantum" ever being used by anybody. Most of us grew out of it, but not the people in this film. Here is another sampling to go with the ones at the head of the chapter.

> "To acknowledge the quantum self, is to acknowledge the place where you rarely have chosen to acknowledge mind...when that shift of perspective takes place, we say that somebody has been enlightened."
>
> Amit Goswami, Ph.D[20]

And finally,

> "We're living in a world where all we see is the tip of the iceberg. The immense tip of the quantum mechanical iceberg."
>
> John Hagelin, Ph.D[20]

Well, maybe so.

---

[19]**Capra, 211.**
[20]***Bleeps.***

Chapter 8

L'Envoi

When I started this exploration I had no idea where it would lead; that we would go from Auden to the Dalai Lama and on to communist ideology. The riches of the quantum theory are so vast that I am sure that we are closer to the beginning than to the end. But here are two of my favorite quotations, one from Bohr and the other from Feynman. They are a good note to end on.

BOHR: For those who are not shocked when they first come across quantum theory cannot possibly have understood it.

And

FEYNMAN: Might I say immediately ... we always have had a great deal of difficulty in understanding the world view that quantum mechanics represents ... I cannot define the real problem, therefore I suspect there's not a real problem, but I am not sure there's no real problem.

Chapter 9

Appendix

There are no new discoveries that involve the quantum theory itself. This is unfortunate, because then we might have some experimental guidance as to how to modify the theory. There is some agreement among thoughtful students of the subject that there is no entirely satisfactory interpretation of the quantum theory. This has provoked a thoughtful practitioner like Freeman Dyson to propose that the theory only applies to probabilistic predictions of the future, but that the past is not probabilistic so the quantum theory does not apply to it. It happened and should be described classically. Dyson argues that there is no way in the standard quantum theory to formulate the statement that that atom decayed yesterday at 4 :P.M. I don't think many other physicists are willing to go this far. But I would like here to explain the discontents that many of us feel and which were so well summarized by the late John Bell.

I spent the winter and spring of 1989 at CERN. I had, among other things, the definite intention to do a series of interviews with Bell. I had known him for some twenty years but I did not know much about his personal history. He told me that he came from a very modest Protestant family in Ireland. "Poor but honest, as Bell put it. Under the normal evolution of things he would have been expected to drop out of school at fourteen and begin to go to work to support his family. In Ireland at the time there was no free high school education. But his mother realized that there was something special about Bell and enough money was scraped up to allow him to continue his education. He worked his way through college but soon after he got his undergraduate degree he went to work for the British nuclear energy program there among other things helping to design an accelerator. Even when he went to CERN he always had an interest in accelerator design and he listed himself as a "quantum engineer."

His discontent with the standard formulations of the quantum theory went back to when he first learned it at the university. But at CERN this was not part of his remit so he did not do much about it until he took a sabbatical at Stanford where he first proved his theorem about the distinguishing aspects of the quantum theory.

During my interviews Bell often complained about the state of existing textbooks on quantum theory, none of which in his view discussed the foundations of the theory coherently. I finally asked him why he didn't write his own book. He said that he was stuck on the compatibility of the quantum theory and relativity. I much regret that I did not pursue this with him, although in his essays such as "Against Measurement," he spells out his dilemma very clearly. But I would like to begin by describing what he did not mean.

It is well known that making a quantum theory of gravity has turned out to be very difficult. Léon Rosenfeld who was a longtime associate of Niels Bohr took the position that it was impossible and unnecessary because gravity dealt with the macroscopic while the quantum theory dealt with the microscopic. Most physicists would not go this far and would argue that if quantum theory is the theory of the world then it must include gravity. Moreover, where is the precise line between the microscopic and the macroscopic? How many atoms have to be involved? While Bohr never published a full version of what came to be known as the Copenhagen description of quantum mechanics, disciples like Rosenfeld spread the word. For them there was one version of the quantum theory and this was it. Before I turn to that I would like to explain why gravity has been so resistant to integration with the quantum theory, as opposed to, say, electrodynamics.

Soon after quantum mechanics was created in the 1920s, attempts were made to quantize the electromagnetic field. The photon, for example, could be created or destroyed and the new quantum electrodynamics contained the mathematics for describing this. People began using it to calculate things and among them was the so-called "self energy" of the electron. The theory demanded a parameter for the mass of the electron. What was used initially was the so-called "bare mass." This is the mass the electron would have if all its electromagnetic interactions were turned off. But if this happened the electron could no longer respond to electric and magnetic fields so that you could not design an experiment using these fields to measure the mass. But one could try to use the theory to compute, say, the ratio of these masses. This was done and the ratio turned out to be infinite, an absurd answer. Things remained this way until just before the Second

World War. At this point the Dutch physicist Hendrick Kramers introduced a new idea. He suggested that the bare mass should never appear in the theory. It should be replaced by the observable mass. In this way the infinity result was swept under the rug. If you wanted to calculate the self energy you would not include the infinite term since it had already been taken into account. This procedure came to be known as "renormalization."

Nothing much was done with this during the World War. Physicists had other things to do. But after the War new experiments were done that revealed these electrodynamic effects. It was noted that if you renormalized both the mass and the charge then every term in the perturbation expansion which is used to calculate various physical quantities was rendered finite. A theory like this is called "renormalizable." As for electrodynamics Dyson showed that while the series in the coupling constant $\alpha = e^2/hc \sim 1/132$ can be rendered finite in every order it cannot converge. Nonetheless it produces results of remarkable accuracy. Take the magnetic moment of the electron which has been calculated up to the fifth order in α. This calculation agrees with experiment to parts in a trillion! No doubt if one continued the calculation to higher and higher orders, at some point the series would begin to break down but there is no sign of that as yet. Why not carry out a similar program for gravitation? One can readily write down the Feynman graphs that represent the terms in the expansion but there is an irredeemable difficulty. Every order reveals new types of infinities and no finite number of renormalizations renders all the terms in the series finite. The theory is not renormalizable.

Over the years many attempts have been made to make a quantum theory of gravity. In none of my interviews with Bell did he comment on this activity. Nor do I find any references to it in his published essays. Bell does introduce the notion of "for all practical purposes"—FAPP. FAPP when discussing the foundations of quantum mechanics Bell was persuaded to ignore the role, if any, of gravitation. I would like to begin with Bohr's view about the quantum theory, which I suspect is what most physicists adopt without much critical thought. Bell was especially annoyed by this and referred to Bohr as an "obscurantist."

The key to Bohr's views is his separation of the world into apparatus and the systems that the apparatus does measurements on. The apparatus with its dials and pointers must be, Bohr insisted, classical. There are no Heisenberg uncertainties here. They are in the "systems" which are quantum mechanical. This classical world is pre-existent and is in no way to be derived as some sort of limit of the quantum theory. This of course infuriated Bell since Bohr never was precise about the division. Once again how many atoms? But even if you accept this division there is still a problem. While the apparatus might be classical you must be able to describe how it performs its measurements on the quantum mechanical system. It turns out that this process cannot be described by quantum mechanics, at least of the usual FAPP variety. To understand this we must explore the meaning of the wave function.

When Schrödinger first wrote down his wave equation he thought that the solutions were real waves oscillating in space and time. He thought of them as guide waves which for example guided the motions of an electron in a hydrogen atom. It was soon realized that this interpretation had fatal flaws. While the radial solution representing the ground state wave function does fall off exponentially as the distance from the proton increases, it never vanishes. What this would say is that the electron can be guided off to infinity over the course of time. Of course this does not happen. The way out was suggested by Max Born. Oddly, it took until 1954, nearly three decades later, before he was awarded the Nobel Prize for this work. Born suggested that the wave function represented probabilities and not waves in actual space-time. This resolved another issue which had been a concern. The wave function of a single particle is a function of the three space and one time variable—four-dimensional space. But when more than one particle is involved the dimension of the space-time begins to run away. This is a problem if one thinks of the solutions as representing real waves but the problem disappears if these are waves of probability.

In addition to the wave function a quantum system is characterized by "observables." These can include things such as momentum and angular momentum, They are represented by "operators". For example the momentum is represented by a derivative, These observables can have a set of allowed values and the object of a measurement is to discover which of these values the system actually has. One must be careful here not to imply that the system has one of these values prior to measurement. The physicist John Wheeler liked to give the analogy of the quantum umpire calling balls and strikes. "They ain't nothing until I call them." If one insists that the system had the observed measurement value all along, and we are simply discovering it, one can construct paradoxes. To each of the allowed values, there is an associated wave function φ_a. We can expand the wave function of the system ψ in terms of the φ_a. Thus

$$\psi = \Sigma_a c_a \varphi_{a.}. \tag{9.1}$$

Here the c's are numbers. Their squares give the probability that a measurement will reveal that particular value of the observable. But what happens to the wave function as a result of the measurement? What is proposed is that after the measurement the complete wave function disappears and we are left with that component which reflects the result of the measurement. This is often referred to as the collapse of the wave function. Two objections have been raised about this, one of which I think is more serious than the other.

The less serious objection, as far as I am concerned, is that this interpretive procedure is not derived from the theory but is simply tacked on. This seems to me to be common to all physical theories. For example Newton's law F = ma is basically a set of symbols until we are informed as to what they mean. This interpretation is not "derived". It is simply tacked on by fiat. This is true of the probability interpretation of the wave function often referred to as Born's rule. The real problem is that the quantum mechanical measurement process cannot be described by quantum mechanics, at least of the FAPP variety. Why is this? Recall that before the measurement the wave function of the system is given by $\psi = \Sigma_a c_a \varphi_{a.}$. The measurement projects out one component of this expansion, say, φ_a. This projection is irreversible in the sense that given the projected wave function one cannot reconstruct what the prior wave function was. This violates the standard time evolution in the quantum theory. The Schrödinger wave function evolves in time in such a way that at any point in time one can reverse the time sense and recover the original wave function. In particular, the wave function preserves its "norm" during this evolution whereas this collapse does not.

This issue seems to have been understood by people like Bohr. Indeed the mathematician John Von Neumann actually produced a model for how the wave function might collapse in a particular instance. This required a modification of the usual quantum mechanical rules and was a precursor to one of the methods—to be discussed shortly—which has been proposed more recently. But since none of this had any effect on how people actually carried out their experiments there was not much interest. An exception was David Bohm whose mechanics, as I will also discuss shortly, dealt with the issue by removing the distinction between measurements and other sorts of interactions. The wave function plays an entirely different role and there is no collapse. I was learning quantum mechanics at this time, especially in a course I was taking with Julian Schwinger. I tried to read Bohm's papers and at the time they seemed rather complicated to me. I remember saying this to Schwinger and his noting that they seemed too simple to him. I don't know how he reacted to the next development in this area—the "Many Worlds" interpretation of the quantum theory by Hugh Everett the third.

Everett came to Princeton in 1953 as a graduate student in mathematics but switched to theoretical physics. He took an introductory course in quantum mechanics and learned the standard lore about the wave function collapse. Something struck him about the wasteful nature of this and he asked himself why the rest of the wave function has to get lost. Why can't it persevere, but in a domain inaccessible to the original observer? He was fortunate to find a thesis advisor, John Wheeler, who had a remarkable tolerance for *outré* ideas. Everett got his degree in 1956. He produced a paper called "The Theory of the Universal Wave Function," which was not published until 1973. In the meanwhile the physicist Bryce De Witt had learned about it and in 1970 published an article in the trade journal *Physics Today* in which he described it as the "many worlds" interpretation—which had a nice science fiction ring to it—and which has been its appellation ever since. In 1956 there was still military conscription in the United States, and physicist or not you could get drafted into the Army. To avoid this Everett became an analyst for military strategies in various companies, some of which he founded. He did keep a bit of a hand in quantum theory, and in 1959 Wheeler arranged for Everett to have a visit to Copenhagen to try out his ideas on Bohr and company. It was a disaster. Bohr had no interest and Rosenfeld called him an idiot. After this Everett rarely spoke about his quantum mechanical ideas in public. He died suddenly at the age of 51 on July 19th, 1982.

Bell of course studied Everett's ideas and frequently commented on them in his essays. He had a more sympathetic view than the Copenhagen people, although he had misgivings. He saw in the many worlds an echo of the one interpretation which Bell liked the best—the de Broglie–Bohm pilot wave, which I will come to shortly. But he found the many worlds interpretation flawed. In the first place it was solipsistic in the extreme. Some experimenter working in some obscure laboratory somewhere had the ability to create new and unobservable universes to him or her and these kept propagating off in space-time to perhaps be observed somewhere by someone else with no awareness of their provenance. The whole thing seemed to Bell like an unnecessary extravagance. There were also issues with relativity which I wiil come back to later. On the plus side it did inspire a new generation of physicists to think about the foundations of quantum theory, something which their teachers had decided was done and dusted.

Before turning to Bell's favorite I will say a few brief words about the others.

John Bell died unexpectedly on October 10th, 1990 at the age of sixty two. I do not think he would have seen the work of Murray Gell-Mann and James Hartle but I suspect that he might have had similar reservations about it that he did express about the "many worlds." The development that led to this work began with a paper of Dirac which he published in 1933. It was pretty much ignored until Feynman took it up for a thesis that he did under the supervision of Wheeler. It was finished in 1942. Feynman then went off to Los Alamos so he did not publish anything about it until 1946. The essential idea is this. Suppose at some time t, within the limits of the uncertainty principles, measurements have been performed on such things as the position and momentum, and perhaps the spin, of some particle. The particle then evolves in time while undergoing interactions with various forces. At a later time these same quantities are measured. If this was classical physics we could predict with certainty the results of these measurements. In quantum mechanics all we can do is predict probable values. Indeed in the usual interpretation of the theory, none of these properties have any value until a measurement is made. In classical physics such a particle would trace out an orbit and we could be certain that there was such an orbit even though we do not choose to observe it at every point. In the usual interpretation of the quantum theory, we have no right to assume anything about orbits in the absence of measurement. This means that the particle could have had several possible "paths" or "histories" between the times that we did our observations. In computing the probabilities we have to sum over all these histories. This being quantum mechanics these histories can interfere, become "entangled" with each other the way that waves do. But interactions can cause these histories to "decohere." Indeed under the right circumstances these decoherent histories behave at least approximately like classical orbits. This is how, according to this picture, the classical world emerges from the quantum theory. In this picture, wave functions don't collapse. They branch but we, the observer, follow only one branch while the others simply remain unobserved.

I would imagine that Bell would have objected to this solipsism. Once again an observer somewhere has produced an alternate history of the world. Moreover we have all these discarded histories cluttering up space-time. I don't think that the elaborate and impressive formalism that Gell-Mann and Hartle have produced would have changed Bell's mind.

Some years ago I heard Eugene Wigner lecture about the measurement problem. It was surprising to me to hear someone of his generation admit that there was a problem and indeed, that it was the same problem that I recognized—the collapse of the wave function. I must admit that I was somewhat disappointed with Wigner's "solution." He wanted to make a small addition to the Schrödinger wave equation that would produce the collapse. It would have no other observable effects. This seemed pretty *ad hoc* to me.

A related but different version of this idea was proposed by the physicists G.C. Ghiradi, A. Rimini and T. Weber—GRW. Bell did comment on this. In the GRW version, as in Wigner's proposal, there is something added to the Schrödinger equation that makes the wave function collapse. But in this version this addition is transitory. It comes on for small instants of time at random and then shuts off until it again comes on. If the constants are adjusted appropriately it will come on when needed and not come on when not needed. Bell did not seem to have anything against this idea although it was not his favorite. His favorite is what he always referred to as the de Broglie–Bohm pilot wave. The addition of de Broglie's name always seemed to me a bit much. We do owe to de Broglie the introduction of the wave nature of matter. His waves were pilot waves but he never had an equation for them. That was the work of Schrödinger. De Broglie's efforts were strongly criticized by Pauli and he gave up until Bohm began publishing his work and de Broglie began claiming priority. I have attempted to read both of their publications, and Bohm, as far as I am concerned, is in a totally different category—far more comprehensible. I always refer to this work as "Bohmian mechanics" which is something Bohm did not do. In Bohmian mechanics there is a Schrödinger wave equation and a wave function determined by it. But the observables are classical particles with real Newtonian-like trajectories. These trajectories follow a force law with a force that has two sources. One source is just whatever forces the Newtonian particle would have in the absence of the quantum theory. But there is a second source—a so-called quantum potential—which is determined by the wave function. This can be worked out so that the results are identical to what you would find in ordinary quantum theory. There are no special operations for measurements. The classical particles are simply guided to reproduce whatever the quantum measurements would predict. The wave function never collapses and simply continues to determine the quantum potential. This is not the place to try to explain how the uncertainty principles emerge nor how particles with properties like spin are treated.

The bottom line is that the results of the theory are identical to ordinary non-relativistic quantum mechanics. This last is the rub. Bell showed that if you have interacting particles, then the results are hopelessly non-local. You must know instantaneously the influences of the various particles on each other and this involves superluminal signals. Bell never could figure out how to make this theory relativistic, so he never wrote his quantum text. Indeed all of this leads one to conclude that at the present time there is no satisfactory interpretation of the quantum theory even though it is the best physical theory ever produced. We wait for experimental guidance. It is clear that the instantaneous collapse of the wave function is not relativistically covariant. This matters if you think that the wave function represents something more than a kind of notepad for recording probabilities.

In the 1930s von Neumann decided that relativity and conventional quantum mechanics were incompatible. He noted that in conventional quantum mechanics, the positions are operators while the time is not. Much effort, all unsuccessful, was expended to try to find an operator associated with time. Now most physicists believe that this sort of thing is futile. What is relevant is the quantum theory of fields where space and time are parameters. In short, ordinary quantum mechanics cannot be made compatible with relativity and attempts to do so are misguided. Indeed, there is a deeper issue. When the energies of, say, collisions become relativistic, then pairs of particles are produced out of the vacuum. To describe this requires a quantum field theory, and hence relativity and ordinary quantum mechanics are truly incompatible.[1]

Post Script:

P.A.M. Dirac divided quantum mechanical problems into two classes. The second class problems were those that could be solved in terms of the existing quantum mechanics. The first class problems were those that would have to wait for an evolution of the theory. These included the foundations. He once gave an evaluation of his own great text on the theory. He said that he thought that it was a good book but that it was missing a first chapter.

[1] **A reader who would like a more technical discussion of these ideas might consult an article of mine in Volume 4, Issue 2 of *Inference*.**